Springer
Proceedings in Physics 85

Springer
Berlin
Heidelberg
New York
Barcelona
Hong Kong
London
Milan
Paris
Singapore
Tokyo

Springer Proceedings in Physics

48 *Many-Atom Interactions in Solids*
Editors: R. M. Nieminen, M. J. Puska, and M. J. Manninen

49 *Ultrafast Phenomena in Spectroscopy*
Editors: E. Klose and B. Wilhelmi

50 *Magnetic Properties of Low-Dimensional Systems II: New Developments*
Editors: L. M. Falicov, F. Mejía-Lira, and J. L. Morán-López

51 *The Physics and Chemistry of Organic Superconductors*
Editors: G. Saito and S. Kagoshima

52 *Dynamics and Patterns in Complex Fluids: New Aspects of the Physics–Chemistry Interface*
Editors: A. Onuki and K. Kawasaki

53 *Computer Simulation Studies in Condensed-Matter Physics III*
Editors: D. P. Landau, K. K. Mon, and H.-B. Schüttler

54 *Polycrystalline Semiconductors II*
Editors: J. H. Werner and H. P. Strunk

55 *Nonlinear Dynamics and Quantum Phenomena in Optical Systems*
Editors: R. Vilaseca and R. Corbalán

56 *Amorphous and Crystalline Silicon Carbide III, and Other Group IV-IV Materials*
Editors: G. L. Harris, M. G. Spencer, and C. Y. Yang

57 *Evolutionary Trends in the Physical Sciences*
Editors: M. Suzuki and R. Kubo

58 *New Trends in Nuclear Collective Dynamics*
Editors: Y. Abe, H. Horiuchi, and K. Matsuyanagi

59 *Exotic Atoms in Condensed Matter*
Editors: G. Benedek and H. Schneuwly

60 *The Physics and Chemistry of Oxide Superconductors*
Editors: Y. Iye and H. Yasuoka

61 *Surface X-Ray and Neutron Scattering*
Editors: H. Zabel and I. K. Robinson

62 *Surface Science: Lectures on Basic Concepts and Applications*
Editors: F. A. Ponce and M. Cardona

63 *Coherent Raman Spectroscopy: Recent Advances*
Editors: G. Marowsky and V. V. Smirnov

64 *Superconducting Devices and Their Applications*
Editors: H. Koch and H. Lübbig

65 *Present and Future of High-Energy Physics*
Editors. K.-I. Aoki and M. Kobayashi

66 *The Structure and Conformation of Amphiphilic Membranes*
Editors: R. Lipowsky, D. Richter, and K. Kremer

67 *Nonlinearity with Disorder*
Editors: F. Abdullaev, A. R. Bishop, and S. Pnevmatikos

68 *Time-Resolved Vibrational Spectroscopy V*
Editor: H. Takahashi

69 *Evolution of Dynamical Structures in Complex Systems*
Editors: R. Friedrich and A. Wunderlin

70 *Computational Approaches in Condensed-Matter Physics*
Editors: S. Miyashita, M. Imada, and H. Takayama

71 *Amorphous and Crystalline Silicon Carbide IV*
Editors: C. Y. Yang, M. M. Rahman, and G. L. Harris

72 *Computer Simulation Studies in Condensed-Matter Physics IV*
Editors: D. P. Landau, K. K. Mon, and H.-B. Schüttler

73 *Surface Science: Principles and Applications*
Editors: R. F. Howe, R. N. Lamb, and K. Wandelt

74 *Time-Resolved Vibrational Spectroscopy VI*
Editors: A. Lau, F. Siebert, and W. Werncke

75 *Computer Simulation Studies in Condensed-Matter Physics V*
Editors: D. P. Landau, K. K. Mon, and H.-B. Schüttler

76 *Computer Simulation Studies in Condensed-Matter Physics VI*
Editors: D. P. Landau, K. K. Mon, and H.-B. Schüttler

77 *Quantum Optics VI*
Editors: D. F. Walls and J. D. Harvey

78 *Computer Simulation Studies in Condensed-Matter Physics VII*
Editors: D. P. Landau, K. K. Mon, and H.-B. Schüttler

79 *Nonlinear Dynamics and Pattern Formation in Semiconductors and Devices*
Editor: F.-J. Niedernostheide

80 *Computer Simulation Studies in Condensed-Matter Physics VIII*
Editors: D. P. Landau, K. K. Mon, and H.-B. Schüttler

81 *Materials and Measurements in Molecular Electronics*
Editors: K. Kajimura and S. Kuroda

82 *Computer Simulation Studies in Condensed-Matter Physics IX*
Editors: D.P. Landau, K. K. Mon, and H.-B. Schüttler

83 *Computer Simulation Studies in Condensed-Matter Physics X*
Editors: D. P. Landau, K. K. Mon, and H.-B. Schüttler

84 *Computer Simulation Studies in Condensed-Matter Physics XI*
Editors: D. P. Landau and H.-B. Schüttler

85 *Computer Simulation Studies in Condensed-Matter Physics XII*
Editors: D. P. Landau, S. P. Lewis and H.-B. Schüttler

Volumes 1–47 are listed at the end of the book

D. P. Landau S. P. Lewis H.-B. Schüttler (Eds.)

Computer Simulation Studies in Condensed-Matter Physics XII

Proceedings of the Twelfth Workshop
Athens, GA, USA, March 8–12, 1999

With 92 Figures

Springer

Professor David P. Landau, Ph.D.
Professor Steven P. Lewis, Ph.D.
Professor Heinz-Bernd Schüttler, Ph.D.

Center for Simulational Physics
The University of Georgia
Athens, GA 30602-3451, USA

ISSN 0930-8989
ISBN-13:978-3-642-64086-5 Springer-Verlag Berlin Heidelberg New York

Cataloging-in-Publication Data applied for.

Die Deutsche Bibliothek - CIP-Einheitsaufnahme

Computer simulation studies in condensed matter physics :
proceedings of the ... workshop ... - Berlin ; Heidelberg ; New York ;
Barcelona ; Hong Kong ; London ; Milan ; Paris ; Singapore ; Tokyo
: Springer
12. Athens, GA, USA, March 8-12, 1999. - 1999
(Springer proceedings in physics ; 85)
ISBN-13:978-3-642-64086-5 e-ISBN-13:978-3-642-59689-6
DOI: 10.1007/978-3-642-59689-6

Softcover reprint of the hardcover 1st edition 2000

Typesetting: Camera ready copy from the authors/editors
SPIN: 10742206 57/3144 - 5 4 3 2 1 0 - Printed on acid-free paper

Preface

More than a decade ago, because of the phenomenal growth in the power of computer simulations, The University of Georgia formed the first institutional unit devoted to the use of simulations in research and teaching: The Center for Simulational Physics. As the simulations community expanded further, we sensed a need for a meeting place for both experienced simulators and neophytes to discuss new techniques and recent results in an environment which promoted extended discussion. As a consequence, the Center for Simulational Physics established an annual workshop on Recent Developments in Computer Simulation Studies in Condensed Matter Physics. This year's workshop was the twelfth in this series. It was held at The University of Georgia, March 8–12, 1999 as an unofficial satellite conference to the Centennial Meeting of the American Physical Society in Atlanta, GA. The continued interest shown by the scientific community demonstrates quite clearly the useful purpose which the series has served. These proceedings provide a "status report" on a number of important topics. This volume is published with the goal of timely dissemination of the material to a wider audience.

We wish to offer special thanks to IBM Corporation for their generous support of this year's workshop.

This volume contains both invited papers and contributed presentations on problems in both classical and quantum condensed matter physics. We hope that each reader will benefit from specialized results as well as profit from exposure to new algorithms, methods of analysis, and conceptual developments.

Athens, GA, USA,
May 1999

D. P. Landau
S. P. Lewis
H.-B. Schüttler

Contents

Computer Simulation Studies in Condensed Matter Physics: An Introduction
D. P. Landau, S. P. Lewis, H.-B. Schüttler 1

Part I Electronic Structure and Quantum Monte Carlo

Beyond Wavelets: Exactness Theorems for Physical Calculations
T.A. Arias, T.D. Engeness 7
1 Introduction 7
2 Electronic Structure of Matter 8
3 Compact Representation of Fields 10
4 Expression of Physical Couplings 12
5 Efficient Solutions 18
References 21

Screening of a Point Charge: A Fixed-Node Diffusion Monte Carlo Study
Erik Koch, Olle Gunnarsson, Richard M. Martin 22
1 Introduction 22
2 Model Hamiltonian 23
3 Screening of a Point Charge 25
4 Quantum Monte Carlo 28
References 34

Quantum Simulations of Strongly Correlated Electron Systems
S. Zhang, E. C. Allman 37
1 Introduction 37
2 New Quantum Monte Carlo Methods 39
3 Is the Hubbard Model the Right One for High-T_c? 43
4 Concluding Remarks 51
References 51

Competing Effects in Perovskite Manganites: Ferromagnetism vs. Phase Separation
S. Yunoki, A. Moreo, E. Dagotto ... 53
1 Introduction ... 53
2 One-Orbital Ferromagnetic Kondo Model ... 54
3 Two-Orbital FM Kondo Model with Jahn-Teller Phones ... 60
4 Conclusions ... 65
Acknowledgments ... 66
References ... 66

Part II Phase Transitions and Critical Phenomena

Critical Finite-Size Scaling with Constraints: Fisher Renormalization Revisited
M. Krech ... 71
1 Introduction ... 71
2 Model and Simulation Method ... 73
3 Ising Universality Class ... 76
4 XY Universality Class ... 80
5 Summary and Conclusions ... 84
References ... 85

Monte Carlo Simulation of Spin Models with Long-Range Interactions
E. Luijten ... 86
1 Introduction ... 86
2 Description of the Algorithm ... 87
3 Applications ... 91
4 Outlook and Conclusion ... 98
References ... 99

Phase Diagram of the Spin-3/2 Blume-Capel Model
J. A. Plascak, D. P. Landau ... 100
1 Introduction ... 100
2 Monte Carlo Simulations ... 101
3 Summary ... 104
References ... 104

Dynamic Phase Transition and Hysteresis in Kinetic Ising Models
P. A. Rikvold, G. Korniss, C. J. White, M. A. Novotny, S. W. Sides 105
1 Introduction ... 105
2 Model and Relevant Quantities ... 107
3 Decay of Metastable Phases ... 110
4 Results ... 111

5 Summary and Discussion 117
References 118

Application of the 8-State Clock Model to FM/AFM/FM Trilayers
P. D. Scholten, V. Ranjbar 120
References 123

Monte Carlo Study of Critical Point Shifts in Thin Films
O. Dillmann, M. Müller, W. Janke, K. Binder 124
1 Introduction 124
2 Model 124
3 Phase Coexistence Curve and Scaling Predictions 125
4 Monte Carlo Procedures and Results 126
5 Summary and Outlook 128
References 128

Incipient Spanning Clusters in Square and Cubic Percolation
L. N. Shchur 129
1 Introduction 129
2 Percolation 130
3 Incipient Spanning Probability 132
4 Coexistence of Incipient Spanning Clusters in 2D 133
5 Computational Method 136
6 Finite Size Corrections of Spanning Probabilities. Free Boundaries . 137
7 Periodic Boundary Conditions 142
8 Incipient Spanning Clusters in Simple Cubic Lattice 143
9 Discussion 144
10 Acknowledgments 145

Part III Spin Glasses

Bivariate Multicanonical Monte Carlo of the 3D $\pm J$ Spin Glass
N. Hatano, J. E. Gubernatis 149
1 Introduction 149
2 Spin Glass 150
3 Multicanonical Monte Carlo Method 154
4 Numerical Results 157
5 Discussions 158
References 160

Monte Carlo Simulation of the Three-Dimensional Ising Spin Glass
M. Palassini, S. Caracciolo 162
1 Introduction 162
2 Model and Simulation 162
3 Finite Size Scaling Analysis 163
4 Nature of the Phase Transition 165
References 166

Part IV Computer Simulation Methods

Molecular Dynamics Simulations with IMD
J. Roth, J. Stadler, M. Brunelli, F. Gähler, J. Hahn, M. Hohl, C. Horn, R. Mikulla, G. Schaaf, H.-R. Trebin 169
1 Introduction 169
2 Online Visualization and the Generation of Pictures 170
3 Metacomputing 171
4 Shock Waves in Quasicrystals and Crystals 172

Further Applications of the Broad Histogram Method for Continuous Systems
J. D. Muñoz, H. J. Herrmann 174
1 Introduction 174
2 The BHMC Method 174
3 Example: The Classical Heisenberg Model 176
4 Results 177
5 How to Include an External Field 177
6 Conclusions 178
7 Acknowledgments 178
References 178

Introduction to the Propp-Wilson Method of Exact Sampling for the Ising Model
M. A. Novotny 179
References 184

Magnetic Properties of Finite Systems: Microcanonical Finite-Size Scaling
M. Promberger, M. Kastner, A. Hüller 185
1 Introduction: Microcanonical Description of Finite Systems 185
2 Microcanonical Finite-Size Scaling (MFSS) 187
3 Conclusion 189
References 190

Molecular Dynamic Simulation of Directional Crystal Growth
B.V. Costa, P.Z. Coura, O.N. Mesquita 191
1 Introduction 191
2 Simulation 191
3 Cellular Instability 194
References 196

Part V Polymers and Surfactants

Escape Transition of a Grafted Polymer Chain
A. Milchev, V. Yamakov, K. Binder 199
1 Introduction 199
2 Theoretical Background 200
3 The Model 204
4 Simulational Results 206
5 Summary 209
References 210

Monte Carlo Simulations of Model Nonionic Surfactants
A.P. Chatterjee and A.Z. Panagiotopoulos 211
1 Introduction 211
2 Models and Methods 212
3 Results 215
4 Discussion 221
References 221

A New Model to Simulate the Growth of Branched Polymers
H. Rêgo, L. Lucena, G.M. Viswanathan 223
1 The Branched Polymer Growth Model 223
2 Self-Organized Criticality in the Growth of Branched Polymers 224
3 The Fixed Number of Tips Model (FNTM) 225
4 Results 225
5 Conclusion 226
References 227

Molecular Dynamics Simulations of the Adsorption of Single Chains on Surfaces
A. Michel, S. Kreitmeier, B. Wunderlich 228
1 Introduction 228
2 Simulation Method and Model 228
3 Preliminary Results and Discussion 231
4 Acknowledgement 231
References 231

List of Contributors

Allman, E.C. 37
Arias,, T.A. 7

Binder, K. 124, 199
Brunelli, M. 169

Caracciolo, S. 162
Chatterjee, A.P. 211
Costa, B.V. 191
Coura, P.Z. 191

Dagotto, E. 53
Dillmann, O. 124

Engeness, T.D. 7

Gähler, F. 169
Gubernatis, J.E. 149
Gunnarsson, O. 22

Hüller, A. 185
Hahn, J. 169
Hatano, N. 149
Herrmann, H.J. 174
Hohl, M. 169
Horn, C. 169

Janke, W. 124

Kastner, M. 185
Koch, E. 22
Korniss, G. 105
Krech, M. 71
Kreitmeier, S. 228

Landau, D.P. 1, 100
Lewis, S.P. 1
Lucena, L. 223
Luijten, E. 86

Müller, M. 124
Martin, R.M. 22
Mesquita, O.N. 191
Michel, A. 228
Mikulla, R. 169
Milchev, A. 199
Moreo, A. 53
Muñoz, J.D. 174

Novotny, M.A. 105, 179

Palassini. M. 162
Panagiotopoulos, A.Z. 211
Plascak, J.A. 100
Promberger, M. 185

Rêgo, H. 223
Ranjbar, V. 120
Rikvold, P.A. 105
Roth, J. 169

Schüttler, H.-B. 1
Schaaf, G. 169
Scholten, P.D. 120
Shchur, L.N. 129
Sides. S.W. 105
Stadler, J. 169

Trebin, H.-R 169

Viswanathan, G.M. 223

White, C.J. 105
Wunderlich, B. 228

Yamakov, V. 199
Yunoki. S. 53

Zhang, S. 37

Computer Simulation Studies in Condensed Matter Physics: An Introduction

D. P. Landau, S. P. Lewis, and H.-B. Schüttler

Center for Simulational Physics, The University of Georgia, Athens, GA 30602-2451, USA

The general area of computer simulation studies in condensed matter physics continues to be a vibrant area of investigation. The "status report" which is contained in this volume is the result of presentations and discussion which took place during the 12th Annual Workshop at the Center for Simulational Physics. The texts of both longer, invited presentations as well as a number of contributed papers are included. The reader will find that the scope of simulational studies is broad and that substantial potential for cross-fertilization of methods between different sub-fields exists.

To open the volume we present a series of papers on electronic structure and quantum Monte Carlo. First, Arias and Engeness discuss a multiresolution approach for computing the electronic structure of real materials. Building on wavelet theory, they introduce the concept of "semicardinality" and show that basic physical couplings can be computed exactly and efficiently using a semicardinal basis. Koch et al. then investigate the static screening of a point charge in a strongly correlated electron system near the Mott transition using the fixed-node diffusion Monte Carlo method on a Hubbard-like Hamiltonian. They discuss the implications of their results for the superconductivity in alkali-doped fullerenes. In the next paper, Zhang and Allman present a finite-temperature generalization of the constrained-path quantum Monte Carlo method for studying strongly correlated electron systems. This approximate approach is free of the well-known Fermion sign problem, and thus extends the capabilities of quantum Monte Carlo methods to significantly lower temperatures and larger system sizes. Lastly, Yunoki et al. report on a Monte Carlo study of the phase diagrams for one- and two-orbital ferromagnetic Kondo models, the latter including Jahn-Teller coupling. These are electronic models for the perovskite manganites, and both exhibit a mixed-phase region separating the hole-doped ferromagnetic-metal and undoped insulator phases.

The next chapter contains a series of papers on phase transitions and critical phenomena. Krech first describes the influence of constraints on the finite size scaling behavior near phase transitions. Monte Carlo data are presented

Springer Proceedings in Physics, Volume 85
Computer Simulation Studies in Condensed-Matter Physics XII
Eds.: D.P. Landau, S.P. Lewis and H.-B. Schüttler

for both Ising and XY-models in three dimensions and the consequent modifications of the scaling arguments are determined. Luijten then considers spin models with long range interactions and studies them with a new, efficient Monte Carlo algorithm. He discusses the application of this approach to several different Ising models which have quite different critical behavior. Next, Plascak and Landau report a Monte Carlo study of the spin-3/2 Blume-Capel model in two dimensions. A multiphase critical point is found at the end of the first-order line. Rikvold et al. turn to the Monte Carlo study of dynamic phase transitions in kinetic Ising models which exhibit hysteresis. As the frequency of an applied oscillating field is varied, the metastable phase decays through the nucleation and growth of many droplets. Extensive Monte Carlo data are analyzed using finite size scaling with the surprising conclusion that the behavior is consistent with the equilibrium model. Scholten and Ranjbar have studied magnetic trilayers consisting of antiferromagnetic 8-state clock model planes sandwiched between ferromagnetic planes. Preliminary phase diagrams are determined for different layer thickness. Next, Dillmann et al. report new Monte Carlo data for capillary condensation in very thin Ising films. Cluster flipping methods were used, with ghost spins coupled to applied magnetic fields, together with reweighting analysis. In the concluding paper of this section, Shchur reports a very detailed Monte Carlo study of the percolation transition in two dimensional strips. Using a very efficient code of the Hoshen-Kopelman type, he searches for incipient spanning clusters for different strip widths. The data confirm the conjectured formulae for the probability exponents.

The third chapter contains two contributions on recent large-scale simulations of the three-dimensional $\pm$ J Ising spin glass model. Hatano and Gubernatis have used a bivariate multicanonical Monte Carlo approach to obtain the transition temperature Tc and the order parameter distribution function P(q). They report that P(q) in the low-T phase exhibits certain features which are consistent with the droplet picture of the Ising spin glass. Palassini and Caracciolo employ Monte Carlo and finite-size scaling techniques to study the phase transition in the same model. Approaching the transition from the paramagnetic side, they demonstrate the existence of a universal finite-size scaling function. Their extrapolated infinite volume data are consistent with both a continuous phase transition and an essential singularity at finite temperature T, but incompatible with an essential singularity at T=0.

The next chapter is devoted to new computer simulation methods and novel applications of existing methods. The paper by Roth et al. describes an extended and improved version of the IMD software package for massively parallel 2D and 3D molecular dynamics simulations. The package allows for the possibility of clustering in a heterogeneous, massively parallel supercomputing environment, and these "metacomputing" capabilities are demonstrated by simulations of shock waves in quasi-crystals. The paper by

Muñoz and Herrmann describes the extension of a new Monte Carlo data analysis technique, the "broad histogram method", to systems with continuous degrees of freedom. The method can be used to extract the density of states from certain microcanonical simulations and allows for an accelerated estimation of thermodynamic quantities such as the specific heat. Novotny's paper describes the recently developed "coupling-from-the-past" method, originally due to Propp and Wilson, which allows one to generate uncorrelated random samples drawn from the exact equilibrium distribution of a certain class of statistical mechanics models. The method is illustrated for the two-dimensional homogeneous field and random field Ising models with 1st-neighbor ferromagnetic interactions. The work by Kastner et al. explores the finite-size scaling approach in the microcanonical ensemble. The relation between canonical and microcanonical finite-size scaling is discussed, and it is argued that the microcanonical approach may exhibit smaller finite-size effects than the canonical approach. Costa et al. describe a molecular dynamics simulation of directional solidification. A Lennard-Jones potential is used and the structure of the solid-liquid interface is examined.

The final chapter is devoted to applications of computer-simulation techniques to the study of polymer and surfactant systems. First, Milchev et al. report on a Monte Carlo simulation of a polymer chain squeezed between a piston and the surface to which it is grafted at one end. They examine a phase transition in which, at a critical piston height, a portion of the "imprisoned" polymer suddenly escapes from the piston region. Their results are then compared to previous theoretical predictions. Next, Chatterjee and Panagiotopoulos perform grand-canonical Monte Carlo simulations with histogram reweighting to study the thermodynamics of micelle formation for a lattice model of an amphiphilic surfactant near the so-called critical micellar concentration. Rêgo et al. propose a model for the study of branched polymers for which the number of active tips is fixed with time. Results of Monte Carlo simulations of this model are compared with those for another model based upon the concept of self-organized criticality. In the final paper of this volume Michel et al. describe molecular dynamics simulations of the adsorption of single-polymer chains on surfaces. The conformation of the resultant structures is then studied as a function of temperature.

Part I

Electronic Structure and Quantum Monte Carlo

Beyond Wavelets: Exactness Theorems for Physical Calculations

T.A. Arias and T.D. Engeness

Department of Physics
Massachusetts Institute of Technology, Cambridge MA 02139, USA

Abstract. This paper develops the use of wavelets as a basis set for the solution of physical problems exhibiting behavior over wide-ranges in length scale. In a simple diagrammatic language, this article reviews both the mathematical underpinnings of wavelet theory and the algorithms behind the fast wavelet transform. This article underscores the fact that traditional wavelet bases are fundamentally ill-suited for physical calculations and shows how to go beyond these limitations by the introduction of the new concept of semicardinality, which allows basic physical couplings to be computed *exactly* from very sparse information, thereby overcoming the limitations of traditional wavelet bases in the treatment of physical problems. The paper then explores the convergence rate of conjugate gradient solution of the Poisson equation in both semicardinal and lifted wavelet bases and shows the first solution of the Kohn-Sham equations using a novel variational principle.

1 Introduction

Problems in the physical sciences often involve behavior spanning many length scales. Three ingredients, which do not necessarily follow one from the other, are needed to deal effectively with such problems : (1) compact representation of the fluctuations of physical fields over different length scales in different regions of space, (2) economical expression of the physical couplings among such fields, and (3) efficient means for solving the resulting equations. The discussion below gives a very powerful general means for dealing with such problems. Although, the discussion is general, for concreteness we shall consider the calculation of electronic structure of matter, reviewed briefly in Sec. 2, as a prototypical example.

As we shall see in Sec. 3, the central concept underlying wavelet theory, multiresolution analysis, is a very elegant and powerful mathematical tool for providing compact representations, but the expression of the most common physical couplings is very awkward in traditional wavelet bases. Sec. 4 shows how the introduction of a new concept, *semicardinality* provides a means of overcoming this limitation and providing an extremely efficient and, surprisingly, *exact* means of expressing the two most fundamental physical couplings. Finally, Sec. 5 shows extremely efficient methods for solving Poisson's equation and the Kohn-Sham equations within this new framework.

Springer Proceedings in Physics, Volume 85
Computer Simulation Studies in Condensed-Matter Physics XII
Eds.: D.P. Landau, S.P. Lewis and H.-B. Schüttler

2 Electronic Structure of Matter

2.1 Multiscale Nature

It is well known that the electronic wave functions in molecular and condensed-matter systems vary much more rapidly near the atomic nuclei than in interatomic regions. In the immediate vicinity of the nucleus and its strong attractive potential, the electrons possess large kinetic energies, as reflected by high spatial frequencies evident in the orbitals. Figure 1 illustrates this behavior, using the carbon atom as an example. The curves in the figure show the Kohn-Sham orbitals of the atom as computed within the local density approximation [1] to density functional theory [2]. The high-frequency "core" region extends only approximately 0.5 bohr radii out from the nucleus, beyond which the variations in the wave functions are quite smooth. Resolving the cusps in the s states of this atom requires a resolution on the order of 0.03 bohr (corresponding to a plane wave cutoff [3] of nearly 160,000 rydberg). To provide this resolution uniformly throughout the computational cell of these calculations, which is 8 bohr on a side, would require a basis with 16 million coefficients. The vast majority of these basis functions would be wasted as they would serve to provide unnecessarily high resolution outside the core region.

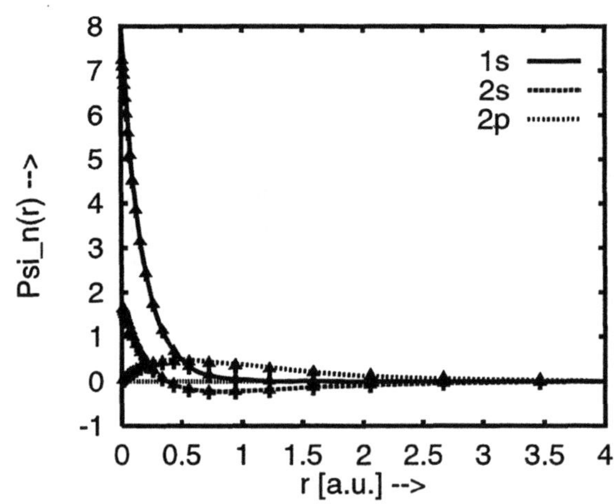

Fig. 1. Kohn-Sham orbitals of the carbon atom within the local density approximation from standard atomic software (*symbols*) and multiresolution analysis(*curves*) [4]

The issue of multiple length-scales in electronic structure is not new. It has driven the development of a variety of techniques which are now quite mature, including the atomic sphere family of approaches, which uses one type of basis set inside of a set of spheres organized around the nuclei and another type of basis set outside of the spheres, and the plane wave pseudopotential approach, which replaces the atomic core with an effective potential manufactured to have similar scattering properties. While each of these approaches has had great success, none is systematically improvable to complete convergence in a simple, practical manner, and each requires great care and expertise in the selection and construction of the atomic spheres or in the development of appropriate pseudopotentials. As a result, a general method is still needed to obtain unambiguous results of sufficient accuracy to permit direct and systematic study of the relative accuracy of competing density functionals and alternate theories of electronic structure. This situation calls for precisely the capabilities of multiresolution analysis, which holds the promise of at last enabling the systematic evaluation of different theories of electronic structure at high precision.

2.2 Electronic Structure

Throughout this paper, we shall work within the local density approximation (LDA) to density functional theory [1]. A novel way to express the Kohn-Sham equations within this approximation, first introduced in [5], is through the following variational principle (in atomic units $\hbar = m = e = 1$),

$$0 = \delta \mathcal{L}_{LDA} = \delta \int \Bigg[\sum_i f_i \frac{1}{2} |\nabla \psi_i(\boldsymbol{r})|^2 + V_{\text{nuc}}(\boldsymbol{r}) n(\boldsymbol{r}) + \epsilon^{\text{xc}}(n(\boldsymbol{r}))n(\boldsymbol{r}) + \phi(\boldsymbol{r})n(\boldsymbol{r}) - \frac{1}{8\pi}|\nabla\phi(\boldsymbol{r})|^2 \Bigg] d^3\boldsymbol{r}. \tag{1}$$

Here, the minimization is over orthonormal sets of Kohn-Sham orbitals $\{\psi_i(\boldsymbol{r})\}$ and the electrostatic (Hartree) potential $\phi(r)$ of the electrons, $V_{\text{nuc}}(r)$ is the potential of the nuclei, the electron density is defined as

$$n(r) \equiv \sum_i f_i |\psi_i(r)|^2, \tag{2}$$

the f_i are the occupancies of the Kohn-Sham orbitals, and $\epsilon^{\text{xc}}(n)$ is the exchange-correlation energy per electron in a uniform electron gas of density n, a highly non-linear function. The advantage of this new variational principle is that it does not involve the traditional long-range Hartree integral,

$$E_{\text{Hartree}} = \int d^3\boldsymbol{r} d^3\boldsymbol{r}' \frac{n(\boldsymbol{r})n(\boldsymbol{r}')}{|\boldsymbol{r}-\boldsymbol{r}'|}. \tag{3}$$

It thus expresses the physics of electronic structure in terms of two simple types of coupling: (1) purely local, but possibly non-linear couplings, and (2) semi-local quadratic couplings among the fields through differential operators. Any successful approach to electronic structure and other physical problems must specify how to deal effectively with these two forms of coupling. This new variational principle also leads to a novel solution of the Kohn-Sham equations (Sec. 5.2).

3 Compact Representation of Fields

3.1 Two-Scale Decomposition

Figure 2a illustrates the concept of *two-scale decomposition* for functions in one dimension. The first row shows a basis set capable of describing functions which fluctuate on length scales greater than the unit distance. This basis consists of copies of a single function $s(x)$, illustrated as piece-wise linear functions of triangular appearance in the figure, translated onto each integer point n along the real axis. The linear space of functions which may be written in terms of this basis is called V_0. The simplest way to double the resolution of this basis is to simply "compress" it by a factor of two, reducing the horizontal scale of both the functions and of the lattice of points on which they sit, as sketched in the final row of the figure. The space spanned by this finer basis is called V_1. Two-scale decomposition builds upon the original coarser basis for V_0 to produce a basis equivalent to the finer basis for V_1 by adding a new set of basis functions $d(x)$ called *detail functions*, or sometimes *wavelets*. The figure shows these functions along the center row. The *detail space*, denoted W_1, is the space of functions which may be written solely in terms of these detail functions. For the combined basis to be equivalent to the "compressed" basis, every function in V_1 must be expressible as a sum of one function in V_0 and one in W_1. In the language of linear subspaces, $V_0 \oplus W_1 = V_1$.

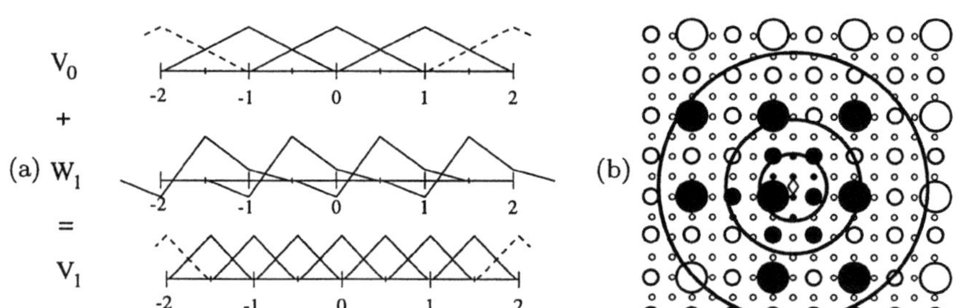

Fig. 2. Two-scale decomposition. (**a**) One dimension. (**b**) Multiple dimensions. Coarse grid (*larger circles*), detail points of finer grids (*smaller circles*), atomic nucleus (*diamond*), spheres of resolution (*large circles*), surviving basis functions (*filled circles*)

This condition determines much about the form of the basis functions. First, the spaces $V_0 \oplus W_1$ and V_1 must have the same dimension, and thus equal numbers of basis functions, so that there must be one detail function for each point which appears in the finer lattice but not in the original coarse lattice. (In the figure, these are the odd half-integer points.) Next the basis functions for both V_0 and W_1 must also be in the space V_1. This specifies a profound restriction on the original basis function, that it may be written exactly in terms of compressed and translated versions of itself,

$$s(x) = \sum_n c_n s(2x - n). \tag{4}$$

This relation which connects the same basis functions on two different scales is known as the *two-scale relation*, the coefficients c_n are known as the *two-scale coefficients*, and the functions satisfying this condition are known as *scaling functions.* (For a detailed discussion, see [6].) Finally, $V_0 \oplus W_1 = V_1$ implies that the detail functions are simply linear combinations of the compressed basis functions,

$$d(x) = \sum_n d_n s(2x - n), \tag{5}$$

where the coefficients d_n are known as the *detail coefficients.*

The preceding conditions ensure that $V_0 \oplus W_1 \subset V_1$. To show that the combined and the finer spaces are indeed equal, one must also show that all functions in V_1 may be written in terms of a linear combination of coarse scaling functions and detail functions. In as much as the two bases contain equal numbers of functions and all functions in either space are linear combinations of finer scaling functions, this amounts to showing that a particular linear system of equations involving the two-scale and detail coefficients always has a solution. There is therefore a certain determinant which must be verified to have non-zero value [7, Theorem 5.16], a condition easily satisfied in practice.

3.2 Multiresolution Analysis

To go beyond two-scale decomposition, we note that starting from V_0, one may produce an entire class of bases of successively higher resolution by reducing the scale of space by successive powers of two. This process defines a sequence of lattices $C_0, \ldots C_N, \ldots$, associated with a sequence of bases of increasing resolution, $V_0, \ldots V_N, \ldots$. Simply changing the horizontal scale in Fig. 2a and the variable x in Eqs. (4,5) by a factor of 2^Q gives the prescription for defining the detail space W_{Q+1} which gives $V_Q + W_{Q+1} = V_{Q+1}$: the basis for W_{Q+1} simply consists of detail functions compressed by a factor of 2^Q, $d(2^Q x)$, centered on the points $D_{Q+1} \equiv C_{Q+1} - C_Q$ which appear in C_{Q+1} but not C_Q. Finally, iterative application of two-scale decomposition can produce spaces V_N of *arbitrarily* fine high resolution, $V_N = V_0 \oplus W_1 \ldots \oplus W_N$, and thus express the entire Hilbert space of functions in one dimension in terms of a basis of hierarchical levels of detail.

These ideas apply equally well in multiple dimensions, as Fig. 2b illustrates. Again, the multiresolution analysis begins with a coarse lattice C_0 on which are centered the coarse scaling functions (large circles in the figure). Then, to double the resolution, detail functions appear on the points of D_1 (intermediate circles in the figure), and to double the resolution yet again, detail functions compressed by a factor of two are added to the points of D_2 (small circles in the figure).

Although such a basis separates functions into contributions on different scales, the final basis still consists of one basis function for each point on the finest scale grid, thus requiring many coefficients to represent physical fields. Figure 2b illustrates how to reduce the size of the basis without significant loss of information. Only the immediate vicinity of the nuclei requires the very highest level of resolution N, and successively lower levels of resolution are required as we move outward from the nuclei. Therefore, about each nucleus we draw a set of successively inscribed spheres of appropriate radii for the scales $0 \ldots N$ chosen to cut off functions with coefficients below a predetermined tolerance, and we keep in the basis only those functions of a given scale which fall within the corresponding sphere (filled circles in the figure). We refer to this process as *restriction*. This process may be carried out adaptively if desired [8].

Such restriction maintains a description equivalent, within the predetermined tolerance, to the full basis V_N and thus the underlying *uniform* basis on the finest scale. In contrast to finite element approaches, the basis functions of a multiresolution analysis constructed this way do not move with the atoms. The only effect of the motion of the atoms is to turn on or turn off basis functions whose coefficients are below the selected tolerance.

Figure 1 shows two sets of results whose comparison demonstrates how well this approach compacts the representation of electronic wave functions. The triangular points in the plot display the results of an essentially exact calculation within the local density approximation, obtained by exploiting the spherical symmetry of the atom. The curves in the figure show that the results obtained from a full, three dimensional calculation using a restricted multiresolution analysis are nearly indistinguishable from the exact results. Yet, rather than requiring a coefficient for each point of the effective grid of 16 million points, the results in the figure come from a restricted basis with six levels of refinement giving a total of only 2,500 functions, compacting the representation by over three orders of magnitude!

4 Expression of Physical Couplings

4.1 Non-Linear, Local Couplings

Figure 3a gives an explicit, diagrammatic representation of the calculation of a non-linear, local coupling among physical fields. Although the procedure is completely general, as an example we consider the computation of the exchange-correlation energy per particle $\epsilon^{\mathrm{xc}}(\boldsymbol{r})$ in terms of the electronic wave functions. The calculation begins at the top of the figure with the multiscale expansion coefficients C_α for the electronic wave functions and ends at the bottom with the corresponding multiscale coefficients for the exchange-correlation field. The circles represent values during the course of the calculation, and the arrows represent mathematical operations which first multiply the value at the base of the arrow by a coefficient specific to the arrow and then accumulate the result onto the value at the head of the arrow. In the first and final rows, the large circles represent the expansion coefficients for the coarse scaling functions in the multiresolution analysis, the intermediate circles are those for the detail functions and the small circles for the finer detail functions.

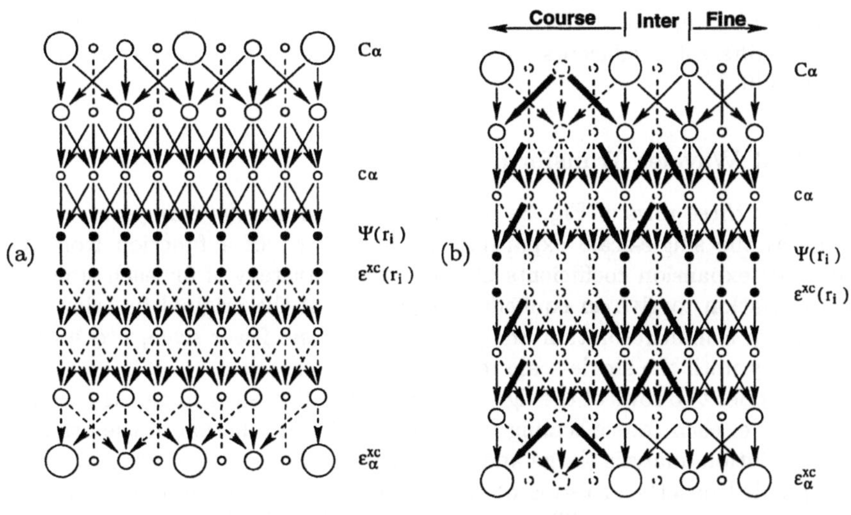

Fig. 3. Calculation of the exchange-correlation field. (**a**) Unrestricted multiresolution analysis. (**b**) Restricted multiresolution analysis. Coarse-scale expansion coefficients (*large circles*), intermediate-scale coefficients (*mid-sized circles*), fine-scale coefficients (*small circles*), function values (*filled points*), multiply-accumulate operations (*arrows*)

The calculation proceeds in three phases. The first, *forward phase* (first through fourth rows of the figure) determines the values of the fields on a set of points in space $\boldsymbol{r}_i$ by first finding the equivalent expansion coefficients c_α for the underlying basis of finest-scale scaling functions and then using these coefficients to compute the real-space values. The first stage of this process (first through third rows) is the fast wavelet transform, explained below in Sec. 4.2. The second stage (third through fourth rows) involves taking the single-scale expansion coefficients c_α (small circles), multiplying by the values of the basis function on the points $\boldsymbol{r}_i$ and accumulating the result onto those points (filled circles), as represented by the arrows. In the second, *coupling* phase (fourth through fifth rows), the known values at the points $\boldsymbol{r}_i$ are combined in the purely local, non-linear fashion prescribed by the coupling. In this example, at each point $\boldsymbol{r}_i$ we take the square magnitude $|\psi(\boldsymbol{r}_i)|^2$ of each wave function, add the results together to compute the charge density $n(\boldsymbol{r}_i)$ and then evaluate the non-linear function $\epsilon^{\text{XC}}(n(\boldsymbol{r}_i))$. The figure represents this independent, point-by-point operation as a set of simple vertical lines. Finally, once the non-linear field is known at all points in space, for the third, *inverse phase* of the calculation (fifth through final row), one inverts the first stage of the procedure and determines the multiscale expansion coefficients for the field in terms of its real-space values by inverting each stage of the forward phase in reverse sequence. Inspection of each of the forward stages reveals that it is a convolution, the inverse of which is another convolution, which may be implemented in the same form but with the arrows potentially carrying different coefficients. (For a more detailed discussion, we refer the reader to [6].)

4.2 Fast Wavelet Transform

The upper three rows of Fig. 3a illustrate the *fast wavelet transform*, which computes the single-scale expansion coefficients c_α of a function from its multiscale expansion coefficients C_α. The basic strategy is for each stage of the calculation to give an equivalent expansion of the original function but in terms of a multiresolution analysis involving one fewer scale, proceeding in this way until the single-scale representation is reached.

The upper two rows of the figure illustrate the first such stage. As Eqs. (4, 5) prescribe, each coarse scaling function and detail function of the original multiresolution analysis (large and intermediate circles of the first row) may be expanded exactly in terms of the intermediate scaling functions (larger circles of the second row). Thus, the arrows which multiply the values carried by the two sets of larger circles on the first row by the corresponding two-scale c_n or detail d_n coefficient and accumulate the result onto the scaling functions of the next scale compute the required equivalent expansion of the original function in a multiresolution analysis of one fewer scale. The coefficients of the finer detail functions play no role in this process, and so are simply copied to the second row as indicated by dashed lines in the figure. Each stage proceeds exactly in this manner.

The number of floating point operations in this transform varies strictly linearly with the number of points in the grid, and thus scales superiorly even to the fast Fourier transform. Typically, the fast wavelet transform requires only approximately 60 floating point operations per grid point. This efficiency, however, is quickly lost when using a restricted multiresolution analysis.

4.3 Impact of Restriction

Figure 3b illustrates the impact of restricting the multiresolution analysis on the calculation of local, non-linear couplings. The basis functions indicated by dashed circles are to be removed from the basis, providing a coarse resolution on the left, an intermediate level of resolution near the center, and a fine resolution on the right. The arrows show the same flow of information as before, but now are drawn to convey the impact of the restriction on the progress of the calculation. Under the restriction, there are three distinct types of information flow. The dashed arrows carry information onto values which have been restricted from the basis, and so we may ignore them in the calculation. The thin, solid arrows carry information from functions in our basis to other functions in our basis and represent calculations which we must perform. Finally, the thick, solid arrows are problematic and carry information from functions restricted from the basis onto function maintained therein and thus have the potential to corrupt the final results of the calculation.

In practice, the thick arrows in the forward phase of the transform (upper portion of the figure) do not corrupt the calculation significantly because the expansion coefficients at their base were determined to be small when the restriction for the basis was chosen. The difficulty is in lower portions of the figure, where the arrows carry not expansion coefficients, but *real-space values* of the fields, which need not be small. Tracing back the flow from these arrows, it is evident that to properly compute the final coarse expansion coefficients ultimately requires computing the the non-linear coupling on all points of the finest grid.

In terms of computational expense, this sacrifices much of the benefit of the multiresolution analysis. Returning to the example of the calculation of the carbon atom, although the wavelet transforms require only about 60 operations per grid point and only 2,500 basis functions were required in the basis, the need to evaluate fields on the finest scale of 16 million points requires the expenditure of 400,000 floating point operations per basis function. A pseudopotential calculation (at 44 rydberg cutoff) in the same cell also requires approximately 2,500 basis functions, but, through the use of the fast Fourier transform, requires only approximately 400 floating point operations per basis function. Without a major advance, the computational expense of representing the physics of the core in a multiresolution analysis thus would require such a computational cost as to relegate such calculations to a very special niche where systematic results are needed very badly.

Our strategy to make the wavelet transform competitive in *restricted* multiresolution analyses is to find multiresolution analyses which eliminate all information flow along the thick arrows in Fig. 3b. We could then compute the results of the transform by simply ignoring the values associated with the dashed circles in the figure so that the transforms may be computed with only approximately 60 floating point operations per *basis function*, less effort than even the fast Fourier transform. Profoundly, without the thick diagonal arrows carrying information from finer scale functions onto neighboring coarser functions, the results obtained in a restricted multiresolution analysis of finite, variable resolution then would be identical to *exact* results obtained by beginning with full knowledge of the physical fields, transforming onto a grid of arbitrarily fine resolution, computing the interaction, and transforming the result back! We refer to this property as *exactness*.

4.4 Exactness Through Semicardinality

We now show a multiresolution analysis which gives rise to such exact results for non-linear, local couplings. For the scaling functions, we draw inspiration from finite-element theory and consider *cardinal* functions, those with zero value on all grid points except for one, where their value is normalized to unity. As Fig. 4a illustrates, the nodal properties of a single-scale basis of cardinal functions ensure that the expansion coefficients of a function and its real-space values on the grid are equal. For the two-scale relation (Fig. 4b), this implies that the two-scale coefficients are simply the values of the scaling function sampled on a grid of half-spacing so that the two-scale coefficients never carry information from a coarse function onto points associated with other coarse functions, an important condition in obtaining exactness.

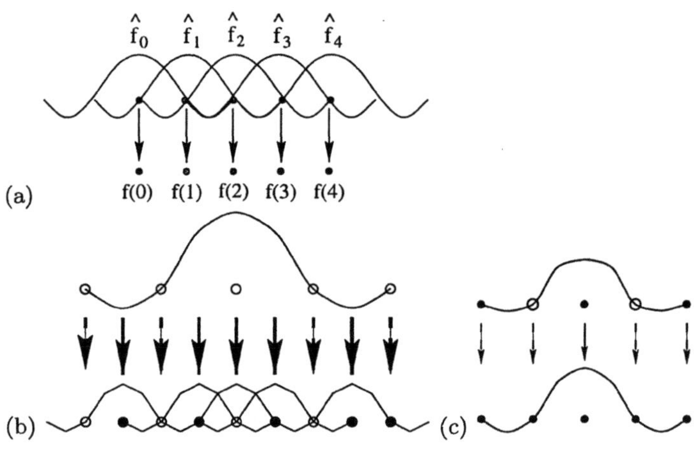

Fig. 4. Semicardinal multiresolution analysis. (**a**) Underlying uniform basis of scaling functions. (**b**) Two-scale decomposition of scaling functions, (c) decomposition of detail functions. Zero coefficients (*dashed arrows*)

For the detail functions, we impose the condition that they be zero on the points of the finer scale whose details they carry (Fig. 4c). One cannot restrict the detail functions to be appear cardinal on the points of the finer scales because then they would become δ-like functions. The coarser-looking, "one-way" form of the cardinality of the basis we construct in this way we refer to as *semicardinality*. As Fig. 4c illustrates, semicardinality implies that all detail coefficients d_n but one are zero, so that the detail functions are simply finer-scale scaling functions, a condition not exhibited in traditional wavelet bases.

An immediate benefit of the cardinality of the scaling functions in a semicardinal basis is that once the forward wavelet transform is complete (Fig. 5a), the expansion coefficients on the finest scale are just the values of the function on the finest grid and the forward phase of the calculation is complete. The nodal conditions in the two-scale and detail coefficient sequences make calculation of the inverse phase of the calculation simple as well. Because at each stage of the forward transform (Fig. 5a) the information flowing out from the scaling functions carries vertically downward without change, in the stages of the inverse transform (Fig. 5a) the final values for the coarser functions may simply be copied again vertically without change. The information flowing out from the detail functions in the forward transform flows vertically, but is met by information flowing diagonally from the scaling functions. Therefore, in the inverse transform, the final values for the detail functions may be copied vertically but met with diagonal information flow of the opposite sign. Note that the pattern of information flow is identical for the corresponding stages of the two transforms. The only change is reversal of the signs of the diagonal arrows.

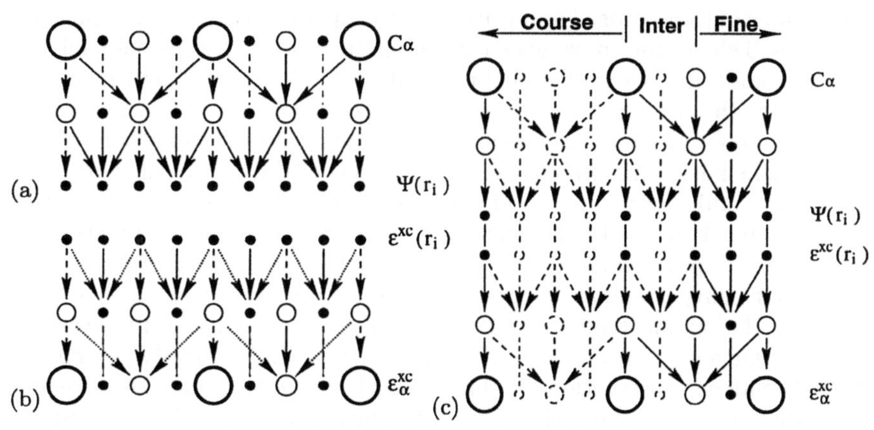

Fig. 5. Calculation of non-linear, local interactions in semicardinal multiresolution analyses. (**a**) Forward transform. (**b**) Inverse transform. (**c**) Complete calculation in restricted basis

Figure 5c shows the computation of a non-linear, local interaction in a restricted *semicardinal* multiresolution analyses. The expression of these interactions involves no corrupting information flow and is now *exact* and may be computed more efficiently than with the fast Fourier transform! Similar results hold in semicardinal bases for the economical expression of couplings through differential operators. One also may compute these interactions *exactly* while working with only those functions that survive a restriction [6].

5 Efficient Solutions

5.1 Poisson's Equation

To solve Poisson's equation, we use the variational principle of minimizing the electrostatic (Hartree) energy. When the potential $\phi(\boldsymbol{r})$ is expanded in a basis, the result is a linear system of equations involving the Galerkin representation for the Laplacian operator. In this section, we focus on the convergence rate of conjugate gradient techniques in solving the resulting equation.

As our example, we compute the electrostatic potential of the carbon atom as a full three-dimensional problem. The basis we employ consists of 12 levels of refinement, each consisting a cubic array of approximately 24^3 detail functions. (We use cubic refinement regions as a matter of computational expediency.) The total number of functions in the basis is approximately $200,000$, and the basis represents a range in length scale and thus condition number of the Laplacian operator in excess of 2^{12} and $2^{24} \approx 16 \times 10^6$, respectively. Under these conditions, one would expect conjugate gradients to require tens of thousands of applications of the operator to solve the equation.

Figure 6 shows the convergence of conjugate gradient methods for solving Poisson's equation with different preconditioners: "diagonal" multiplies the residual by the inverses of the diagonal matrix elements of the Laplacian, "block-diagonal" uses Fourier techniques to invert the diagonal sub-blocks of the Laplacian matrix which connect functions of the same scale with each other, and "multi-level" uses special techniques to invert the upper and lower triangular blocks of the Laplacian matrix connecting functions to functions of only higher or lower scales, respectively. These three sets of calculations were carried out using a semicardinal bases of third order. The data labeled "lift" are results for simple diagonal preconditioning of the Galerkin representation of the Laplacian in a second type of basis, a basis of third-order *lifted wavelets* of the Sweldens construction [9]. The vertical axis of the figure displays the square-magnitude of the residual vector, which varies linearly with the error in the computation of the electrostatic energy because the calculation is variational. To accurately reflect the varying complexity of the different preconditioners, the horizontal axis indicates net computational time in "work units" (WU) defined to be the computational effort of applying the Laplacian operator in the semicardinal basis. For the simple preconditioners, the horizontal axis closely corresponds to number of iterations.

The upshot of the figure is that in these multiresolution bases, conjugate gradients determines the electrostatic energy to within machine precision in only about 100 iterations, three orders of magnitude better than expected!

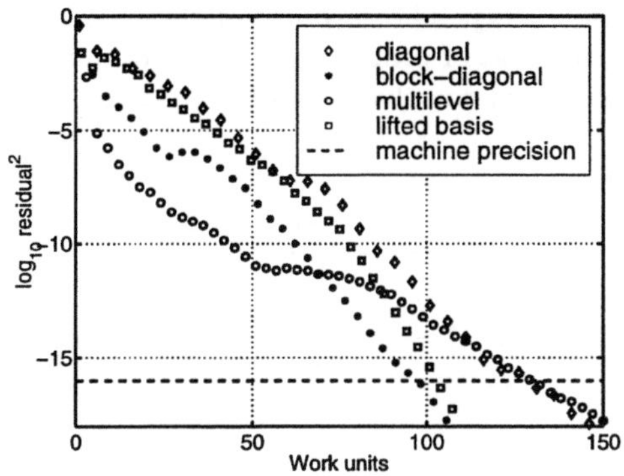

Fig. 6. Solution of Poisson's equation. Convergence of square magnitude of residual vector with various preconditioners in semicardinal and lifted multiresolution bases

(The first reported calculations which uses this basis with the diagonal preconditioner appear in [4].) These very high convergence rates, ~ 0.15 digits/WU, reflect the fact that the semicardinal basis reorganizes the structure of the operator in such a way to enhance its diagonal structure. The most successful approach is the block-diagonal preconditioner. Whereas the multilevel preconditioner gives a dramatic improvement in the number of iterations required to solve the equation, these benefits are outweighed by its computational cost.

In regards to calculation with lifted wavelets, it has been suggested recently that the non-zero integral of the detail functions of a semicardinal basis would create a bottleneck in the solution of Poisson's equation and that, instead, one should use a lifted interpolet basis in which the low order moments of the detail functions vanish [10]. Figure 6b (*squares*) shows that although there is a modest improvement in the convergence at the final stages of the calculation, there is no great advantage in the solution of Poisson's equation from the properties of the lifted basis.

Finally, we consider the impact of problem size on the convergence rate of the electrostatic energy as we increase the level of resolution and the spatial extent of the refinement regions (Figs. 7a and 7b, respectively). Fig. 7a shows that going as far as 24 levels of refinement, the convergence rate of the diagonal and block-diagonal preconditioners does not degrade appreciably despite the condition number of the Laplacian operator now reaching $2^{48} \sim 280$ trillion. This finest level places 1,000 grid points across the extent of the carbon nucleus, more than sufficient to resolve the individual quarks within the nucleons. Fig. 7b explores the behavior of convergence rate as we increase the linear extent of the refinement regions up to about two times the size expected to be relevant in the calculation of electronic structure. Although the diagonal preconditioner degrades somewhat over this range, the block-diagonal conditioner maintains a nearly constant convergence rate of 0.16 digits/WU for the electrostatic energy. The stability of these convergence rates strongly suggests that, regardless of problem size, a fixed number of approximately 100 iterations are required to solve Poisson's equation to a given tolerance.

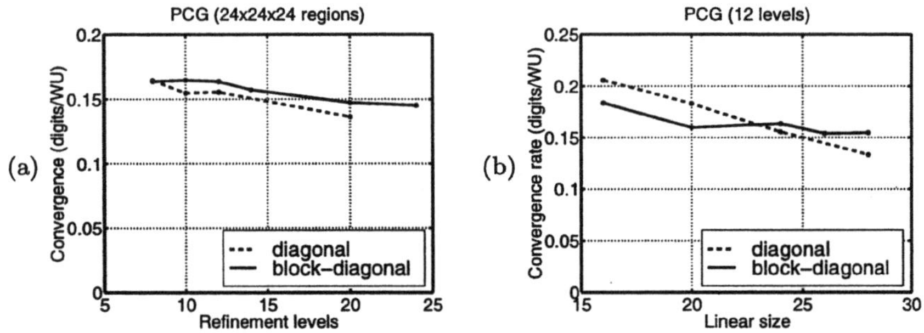

Fig. 7. Behavior of convergence rate in solution of Poisson's equation in semicardinal bases. **(a)** Variation with number of refinement regions, each of size 24^3. **(b)** Variation with spatial extent of refinement regions

5.2 Kohn-Sham Equations

As we saw in the preceding section, the solution of Poisson's equation in semicardinal bases is surprisingly efficient, but still requires on the order of one hundred iterations. This is particularly burdensome in the calculation of electronic structure, where in traditional formulations the Poisson equation must be solved at each step in the minimization of the Kohn-Sham energy functional. We propose instead to search directly for the stationary point of the Lagrangian functional (1). From the signs of the semi-local couplings, it is clear that the stationary point of the Lagrangian functional is a minimum with respect to the electronic wave function fields $\{\psi_i(\boldsymbol{r})\}$ and a maximum with respect to the electrostatic potential field $\phi(\boldsymbol{r})$. Although we are searching for a saddle point, our proposed approach is analogous to iterative conjugate gradient minimization but with a few key differences. Each iteration begins at a state point $(\{\psi_i(\boldsymbol{r})\}, \phi(\boldsymbol{r}))$, but, rather than searching along the downhill-gradient direction $(-\nabla_\psi \mathcal{L},\ -\nabla_\phi \mathcal{L})$, we search along the direction $\boldsymbol{d} \equiv (-\nabla_\psi \mathcal{L}, \nabla_\phi \mathcal{L})$, which is downhill for the electrons and uphill for the electrostatic potential. Next, rather than searching to the point where the functional becomes stationary along the line defined by $\boldsymbol{d}$, we search for the point where the next search direction computed in this same way becomes perpendicular to $\boldsymbol{d}$.

For the beryllium atom, Fig. 8 compares the convergence of this new approach with the traditional approach of applying preconditioned conjugate gradients to minimize the energy functional. The horizontal axis measures

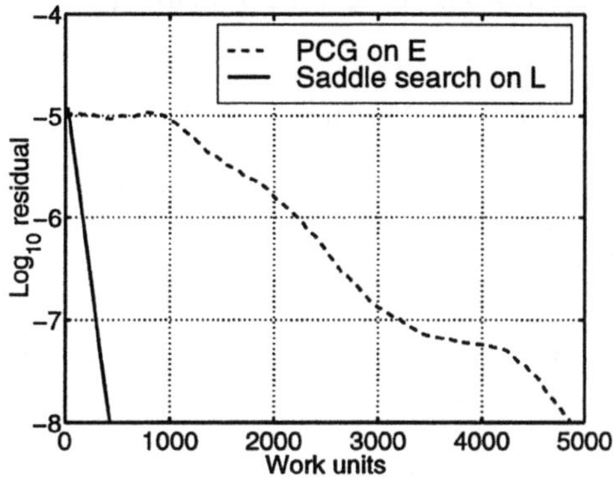

Fig. 8. Comparison of energy and Lagrangian functional solutions to Kohn-Sham equations of density functional theory

computational in work units representing the cost of the application of the Laplacian operator. The figure shows that the computational savings from not having to solve Poisson's equation at each iteration makes the Lagrangian approach a full order of magnitude more efficient than the traditional approach based upon minimization of the energy functional.

References

1. W. Kohn and L. J. Sham. Self-consistent equations including exchange and correlation effects. *Phys. Rev.*, 140:A1133, 1965.
2. P. Hohenberg and W. Kohn. Inhomogeneous electron gas. *Phys. Rev.*, 136:B864, 1964.
3. M.C. Payne, M.P. Teter, D.C. Allan, T.A. Arias, and J.D. Joannopoulos. Iterative minimization techniques for *ab initio* total energy calculations: molecular dynamics and conjugate gradients. *Rev. Mod. Phys.*, 64:1045, 1992.
4. T.A. Arias, K.J. Cho, Pui Lam, and M.P. Teter. Wavelet transform representation of the electronic structure of materials. In Rajiv K. Kalia and Priya Vashishta, editors, *Proceeedings of the '94 Mardi Gras Conference: Toward Teraflop Computing and New Grand Challenge Applications*, page 23, Commack New York, 1995. Nova Science Publishers.
5. R.A. Lippert, T.A. Arias, and A. Edelman. Multiscale computation with interpolating wavelets. *J. Comput. Phys.*, 140:278, 1998.
6. T.A. Arias. Multiresolution analysis of electronic structure: semicardinal and orthogonal wavelet bases. *Rev. Mod. Phys.*, 71:267, 1999.
7. C.K. Chui. *An Introduction to Wavelets.* Academic Press, Boston, 1992.
8. C.J. Tymczak and Xiao-Qian Wang. Orthonormal wavelet bases for quantum molecular dynamics. *Phys. Rev. Lett.*, 78:3654, 1997.
9. W. Sweldens. The lifting scheme: A custom-design construction of biorthogonal wavelets. *Appl. Comput. Harm. Anal.*, 3:186, 1996.
10. S. Goedecker and O.V. Ivanov. Linear scaling solution of the coulomb problem using wavelets. *Solid State Commun.*, 105:665, 1998.

Screening of a Point Charge: A Fixed-Node Diffusion Monte Carlo Study

Erik Koch[1], Olle Gunnarsson[1], and Richard M. Martin[2]

[1] Max-Planck-Institut für Festkörperforschung, 70569 Stuttgart, Germany
[2] Department of Physics and Materials Research Laboratory, University of Illinois, Urbana IL 61801, USA

Abstract. We study the static screening in a Hubbard-like model using fixed-node diffusion Monte Carlo. We find that the random phase approximation is surprisingly accurate even for metallic systems close to the Mott transition. As a specific application we discuss the implications of the efficient screening for the superconductivity in the doped Fullerenes. In the Monte Carlo calculations we use trial functions with two Gutzwiller-type parameters. To deal with such trial functions, we introduce a method for efficiently optimizing the Gutzwiller parameters, both in variational and in fixed-node diffusion Monte Carlo.

1 Introduction

The random phase approximation (RPA) is widely used throughout solid state physics. It properly describes the screening when the kinetic energy is much larger than the Coulomb energy. In the strongly correlated limit, however, it is qualitatively wrong. Little is known for the intermediate regime, where kinetic and Coulomb energy are comparable, and perturbative methods fail since there is no small parameter. In such a situation quantum Monte Carlo is the method of choice. Our goal is to investigate the screening in a strongly correlated system close to the Mott transition. To be specific we focus on the doped Fullerenes, like K_3C_{60}. These materials are strongly correlated and close to a Mott transition [1], but they are also superconductors with quite large transition temperatures ($T_c \approx 30\,\mathrm{K}$ for Rb_3C_{60}) [2]. The superconductivity is driven by the coupling to intramolecular phonons. These phonons mediate an effective electron–electron attraction which is, however, counteracted by the electron–electron repulsion, which is large in such strongly correlated systems. Unlike conventional superconductors, in the doped Fullerenes this repulsion is not reduced much by retardation effects. Therefore efficient screening is important for reducing the electron–electron repulsion sufficiently to allow for an electron–phonon driven superconductivity [3,4]. Using quantum Monte Carlo we find that even for quite strong correlations the RPA gives a surprisingly accurate description of the static screening on the metallic side of a Mott transition until the system is close to the transition [4]. Besides the immediate consequences this result has for our understanding of the superconductivity in the doped Fullerides, it should

Springer Proceedings in Physics, Volume 85
Computer Simulation Studies in Condensed-Matter Physics XII
Eds.: D.P. Landau, S.P. Lewis and H.-B. Schüttler

also have quite general implications for the physics of systems close to a Mott transition.

We start by introducing the model Hamiltonian used to describe the doped Fullerenes. We briefly discuss the Löwdin downfolding technique that gives a general prescription for constructing low-energy Hamiltonians. Next we introduce the screening problem and discuss the results of the Monte Carlo calculations. Finally we review the quantum Monte Carlo methods used. We discuss the choice of the trial wave function and introduce a very efficient method for optimizing the Gutzwiller-type parameters.

2 Model Hamiltonian

Fullerites are crystals made of C_{60} molecules on an fcc lattice. They are characterized by very weak inter-molecular interactions. Therefore the discrete molecular levels merely broaden into narrow, well separated bands [5]. The valence band originates from the lowest unoccupied molecular orbital, which is a 3-fold degenerate t_{1u} orbital. In doped Fullerenes, there are alkali atoms sitting in the space between the C_{60} molecules. They do not affect the band structure around the Fermi energy very much. Only the filling of the t_{1u} band changes, since each alkali atom donates its valence electron. Hence for K_3C_{60} the (3-fold degenerate) t_{1u} band is half-filled. We are mainly interested in the properties of these valence-band electrons. To simplify the description we therefore want to get rid of the other bands. They can be projected out by the Löwdin downfolding technique [6]. The basic idea is to partition the Hilbert space into a subspace that contains the degrees of freedom that we are interested in (in our case the 't_{1u}-subspace') and the rest of the Hilbert space: $\mathcal{H} = \mathcal{H}_0 \oplus \mathcal{H}_1$. We can then write the Hamiltonian of the system as

$$H = \begin{pmatrix} H_{00} & H_{01} \\ H_{10} & H_{11} \end{pmatrix} , \tag{1}$$

where H_{ii} is the projection of the Hamiltonian onto subspace $\mathcal{H}_i$, while the H_{ij} $(i \neq j)$ contain the hybridization matrix elements between the two subspaces. Writing Green's function $G = (E - H)^{-1}$ in the same way, we can calculate the projection of G onto $\mathcal{H}_0$ [7]:

$$G_{00} = \Big(E - \underbrace{[H_{00} + H_{01}\,(E - H_{11})^{-1} H_{10}]}_{H_{\text{eff}}(E)}\Big)^{-1} . \tag{2}$$

We see that the physics of the full system is described by an effective Hamiltonian $H_{\text{eff}}(E)$ that operates on the subspace $\mathcal{H}_0$ only. We have, however, to pay a price for this drastic simplification: the effective Hamiltonian is energy dependent. In practice one approximates it with an energy-independent Hamiltonian $H_{\text{eff}}(E_0)$. This works well if we are only interested in energies close to E_0. In solid C_{60} we have the fortunate situation that the bands retain the character of the molecular orbitals, since the hybridization matrix

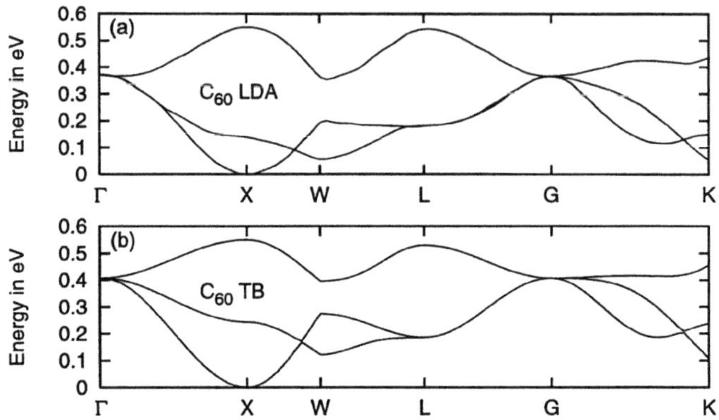

Fig. 1. Band structure (t_{1u} band) of solid C_{60} (fcc) (a) as calculated ab initio using the local density approximation [5] and (b) using a tight-binding Hamiltonian with only t_{1u} orbitals [8]

elements are small compared to the energy separations of the orbitals. In fact we can neglect the other bands altogether and get the hopping matrix elements $t_{in,jn'}$ between the t_{1u} orbitals n and n' on molecules i and j directly from a tight-binding parameterization [8,9]. To demonstrate how well this works, Fig. 1 shows the comparison of the ab initio t_{1u} band structure with the band structure obtained from the tight-binding Hamiltonian with only t_{1u} orbitals.

A realistic description of the electrons in the t_{1u} band also has to include the correlation effects which come from the Coulomb repulsion of electrons in t_{1u} orbitals on the same molecule. The resulting Hamiltonian which describes the interplay of the hopping of electrons and their Coulomb repulsion has the form

$$H = \sum_{\langle ij \rangle} \sum_{nn'\sigma} t_{in,jn'}\, c^{\dagger}_{in\sigma} c_{jn'\sigma} + U \sum_{i} \sum_{(n\sigma)<(n'\sigma')} n_{in\sigma} n_{in'\sigma'} . \tag{3}$$

The on-site Coulomb interaction U can be calculated within density functional theory [10]. It is given by the increase in the energy of the t_{1u} level per electron that is added to one molecule of the system. It is important to avoid double counting in the calculation of U. While the relaxation of the occupied orbitals and the polarization of neighboring molecules have to be included in the calculation, excitations within the t_{1u} band must be excluded, since they are contained explicitly in the Hamiltonian (3). The results are consistent with experimental estimates [11,12]: $U \approx 1.2 - 1.4\,\mathrm{eV}$. For comparison, the width of the t_{1u} band is in the range $W \approx 0.5 - 0.85\ eV$. To properly describe K_3C_{60} the effect of the orientational disorder [8,13] of the C_{60} molecules in the crystal are built into the hopping matrix elements $t_{in,jn'}$. Multiplet effects are not included, since they tend to be counteracted by the Jahn-Teller effect, which is also neglected.

In K_3C_{60} the system has three electrons per molecule. In the limit of weak correlations ($U = 0$), this corresponds to a metal with a half-filled conduction band. In the atomic limit ($U \to \infty$) the Coulomb energy dominates, forcing every molecule to be occupied by exactly three electrons, and suppressing any hopping. This is a Mott insulator. We therefore expect a metal-insulator transition for some finite value of the Coulomb interaction U. For the model Hamiltonian (3) with parameters describing K_3C_{60} it occurs for $U \approx 1.5 - 1.75\,\mathrm{eV}$ [1]. Given the estimates for the true value of U, K_3C_{60} is therefore close to the Mott transition, in a correlated metallic state.

3 Screening of a Point Charge

We now investigate how efficient the screening is in a strongly correlated system like K_3C_{60}. To be specific we analyze how a test charge q sitting on one molecule is screened by the conduction electrons in the t_{1u} band. To describe the influence of the test charge situated on the molecule at site c we include an additional term

$$H_q = qU \sum_{n\sigma} n_{cn\sigma} \tag{4}$$

in the Hamiltonian (3). Determining the electron density at site c for the system without test charge and for the system with a finite q we find the screening

$$\frac{\Delta n}{q} = \frac{n_c(0) - n_c(q)}{q} . \tag{5}$$

Let's first discuss the screening in the RPA. In the random phase approximation it only costs kinetic energy to screen the test charge. In the limit where a typical Coulomb integral U is large compared with the band width W, the kinetic energy cost of screening is relatively small compared with the potential energy gain. Therefore, within the random phase approximation, the screening is very efficient for large U. This means that as the test charge q is introduced, almost the same amount of electronic charge moves away from the site: $\Delta n \approx q$ for large U (see Fig. 2). The random phase approximation neglects, however, that when an electron leaves a site it has to find another site with a missing electron or there is an increase in Coulomb energy of the order of U. Thus the RPA is accurate for small values of U/W, while it is *qualitatively wrong* for large U/W. It is not clear what happens when Coulomb energy U and band width W are comparable.

To address this question we have performed quantum Monte Carlo calculations for the combined Hamiltonian

$$H = \sum_{\langle ij\rangle} \sum_{nn'\sigma} t_{in,jn'} c^\dagger_{in\sigma} c_{jn'\sigma} + U \sum_i \sum_{(n\sigma)<(n'\sigma')} n_{in\sigma} n_{in'\sigma'} + qU \sum_{n\sigma} n_{cn\sigma} \tag{6}$$

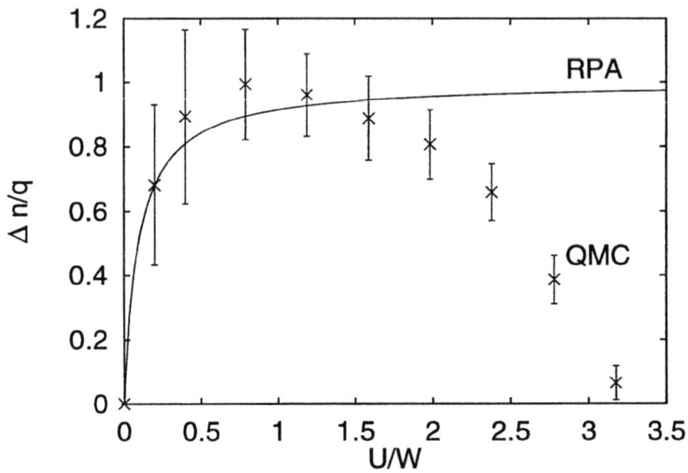

Fig. 2. Screening charge Δn on the site of the test charge ($q = 0.25\,e$) as a function of U/W, extrapolated to infinite cluster size. The full curve shows the screening charge in the RPA, obtained from Hartree calculations for the Hamiltonian (6). The crosses with errorbars give the results of the QMC calculations. The RPA screening remains rather accurate up to $U/W \sim 2$, but fails badly for larger values of U/W. The screening is very efficient for $U/W \sim 0.5 - 2.0$

Since we are interested in the linear response, we should calculate the effect of an infinitesimally small test charge q. Because of the statistical error in a Monte Carlo calculation it is, however, difficult to determine the response to a small perturbation. To get a good signal-to-noise ratio, we would therefore like to use as large a test charge as possible. To estimate how large we can make q and still be in the linear response regime, we have performed Lanczos calculations for a small system of 4 molecules, where exact calculations are possible. Checking the response for different test charges we find that for $q \leq 0.25\,e$ the response is practically linear. The quantum Monte Carlo calculations were then performed for large clusters of N_{mol} =32, 48, 64, 72, and 108 molecules, corresponding to systems with 96, 144, 192, 216, and 324 electrons, and $q = 0.25$. To extrapolate the results $\Delta n(N_{\mathrm{mol}})$ to infinite systems size we used a finite-size scaling of the form $\Delta n = \Delta n(N_{\mathrm{mol}}) + \alpha/N_{\mathrm{mol}}$. The results are shown in Fig. 2. For rather small values of U/W ($\sim 0.5 - 1.0$), the RPA somewhat underestimates the screening. Such a behavior is also found in the electron gas [14]. For intermediate values of U/W ($\sim 1.0 - 2.0$) the RPA gives surprisingly accurate results. For larger U/W the screening rapidly breaks down, and the RPA becomes qualitatively wrong, as discussed above.

The efficient screening almost up to the Mott transition which occurs for $U/W \sim 2.5$ [1] has profound implications for the superconductivity in the alkali doped Fullerenes. We will only give qualitative arguments here, a more detailed discussion can be found in [4]. In BCS theory the superconducting transition temperature is given by

$$T_c \propto e^{-1/(\lambda-\mu^*)} , \tag{7}$$

where $\lambda = N(0)\,V$ describes the electron–phonon coupling, which mediates the effective electron–electron attraction, $\mu^* = N(0)\,U^*$ is the Coulomb pseudopotential that describes the repulsive Coulomb interaction between electrons, and $N(0)$ is the density of states at the Fermi energy. Since the electrons couple to intramolecular phonons, the coupling constant V is to a good approximation a molecular property. Therefore increasing the density of states $N(0)$ will increase the electron–phonon coupling λ. This can be achieved by increasing the lattice constant of the solid Fullerene. Taking the C_{60} molecules further apart will decrease the hopping matrix elements $t_{in,jn'}$ thus narrowing the t_{1u} band and correspondingly increasing the density of states. If also U^* was a molecular quantity, i.e. independent of the lattice constant, and if $V > U^*$ which, of course, is the prerequisite for superconductivity, then according to (7) increasing the lattice constant would raise T_c. Such a variation of the transition temperature with the lattice constant is indeed found experimentally [2]. Here the variation of the lattice constant a is achieved by applying hydrostatic pressure (to reduce a) or by using 'larger' alkali metals like Rb and Cs to increase the lattice constant compared to K_3C_{60} (chemical pressure). This suggests that by inserting even more bulky ions or molecules, like NH_3, into the doped Fullerenes, could be increase T_c even further. As it turns out, however, T_c rapidly *decreases* with a when the lattice constant becomes too large, and eventually superconductivity disappears. This drop in T_c can be understood as a natural consequence of the breakdown of the screening close to the Mott transition. In the doped Fullerenes U^* is essentially given by $U\,(1 - \Delta n/q)$, where U is the unscreened Coulomb matrix element (cf. the interaction term in the Hamiltonian) and $\Delta n/q$ describes the screening by the t_{1u} electrons. U is practically independent of the lattice constant, while the screening efficiency changes with the band width as shown in Fig. 2. For U/W in the region $1.0 - 2.0$ the screening is almost RPA-like and does not vary very much, which means that in this region U^* is almost independent of the lattice constant a. In this region T_c thus increases with a. For even smaller band width, or correspondingly larger lattice constants, the screening rapidly becomes inefficient. Now U^* increases with the lattice constant, leading to a decrease of $\lambda - \mu^*$. Hence for a too large, T_c rapidly decreases with a and eventually vanishes. Since the screening only breaks down close to the metal-insulator transition, we have the interesting situation that *T_c peaks close to a Mott transition!*

The efficient screening found in the Monte Carlo calculations also explains why the coupling to the alkali-phonons is weak. Each C_{60} molecule is surrounded by 14 alkali ions with very weak force constants. When an electron arrives on a molecule one would therefore expect that the surrounding alkali ions respond strongly. This was, however, not confirmed by experiment [15]. That result can be naturally understood as an effect of the efficient screening: When an electron arrives on a molecule other electrons leave, effectively leaving the molecule almost neutral. The alkali ions then only see a small change in the net charge and therefore respond weakly, leading to a small electron–phonon coupling. But being molecular crystals, doped Fullerenes also have intramolecular phonons. Some of those intramolecular phonons shift the t_{1u} levels in such a way that the center of gravity of the energy levels is not changed. These are the modes that are not screened by the transfer of charged. They are therefore the modes that drive superconductivity in the Fullerides.

4 Quantum Monte Carlo

We now turn to the question of how the results shown in Fig 2 were obtained. To keep the notation simple we will discuss the different methods for a simple Hubbard model (only one orbital per site, next neighbor hopping matrix elements $t_{ij} = -t$):

$$H = -t \sum c_i^\dagger c_j + U \sum n_{i\uparrow} n_{i\downarrow} \,. \tag{8}$$

The generalization to Hamiltonians like (3) is straightforward.

The first step in the quantum Monte Carlo approach is to identify a trial function Ψ_T. For the Hubbard model that function should balance the opposing tendencies of the hopping term and the interaction: Without interaction (i.e. for $U = 0$) the ground state of the Hamiltonian (8) is the Slater determinant Φ that maximizes the kinetic energy. Without hopping ($t = 0$) the interaction is minimized. Since only doubly occupied sites, i.e. sites with $n_{i\uparrow} = 1$ and $n_{i\downarrow} = 1$, contribute to the Coulomb energy, the electrons are distributed as uniformly as possible over the lattice to minimize the number of double occupancies. A good compromise between these two extremes is to start from the non-interacting wavefunction Φ but reduce the weight of configurations R with large double occupancies $D(R)$. This leads (up to normalization) to the Gutzwiller wavefunction [16]:

$$\Psi_T(R) = g^{D(R)}\, \Phi(R), \tag{9}$$

with $g \in (0,1]$ the Gutzwiller parameter. In configuration space the Coulomb term in the Hamiltonian is given by $U\,D(R)$. Thus the Gutzwiller factor reflects the interaction term in the Hubbard Hamiltonian. In this spirit we can also construct trial functions for Hamiltonians with additional terms like the screening Hamiltonian (6). We add a second Gutzwiller factor that reflects the interaction with the test charge $q\,U\,n_c(R)$:

$$\Psi_T(R) = g^{D(R)}\, h^{n_c(R)}\, \Phi(R) \,. \tag{10}$$

Changing the additional Gutzwiller factor h we can vary the occupation of the site with the test charge.

4.1 Variational Monte Carlo

Since the Gutzwiller factor, like the interaction term, is diagonal in configuration space, we have to perform a sum over all configurations R in order to calculate the energy expectation value for a Gutzwiller wavefunction:

$$E_T = \frac{\langle \Psi_T | H | \Psi_T \rangle}{\langle \Psi_T | \Psi_T \rangle} = \frac{\sum_R E_{\text{loc}}(R)\, \Psi_T^2(R)}{\sum_R \Psi_T^2(R)} , \tag{11}$$

where we have introduced the local energy for a configuration R

$$E_{\text{loc}}(R) = \sum_{R'} \frac{\langle \Psi_T | R' \rangle \langle R' | H | R \rangle}{\langle \Psi_T | R \rangle} = \sum_{R'}{}' \, t \, \frac{\Psi_T(R')}{\Psi_T(R)} + U\, D(R) \, . \tag{12}$$

Since the number of configurations R grows exponentially with system-size, the summations in (11) can only be done explicitly for very small systems. For larger problems we use variational Monte Carlo [17]. The idea is to perform a random walk in the space of configurations, with transition probabilities $p(R \to R')$ chosen such that the configurations R_{VMC} in the random walk have the probability distribution function $\Psi_T^2(R)$. Then

$$E_{\text{VMC}} = \frac{\sum_{R_{\text{VMC}}} E_{\text{loc}}(R_{\text{VMC}})}{\sum_{R_{\text{VMC}}} 1} \approx \frac{\sum_R E_{\text{loc}}(R)\, \Psi_T^2(R)}{\sum_R \Psi_T^2(R)} = E_T . \tag{13}$$

The transition probabilities can be determined from detailed balance

$$\Psi_T^2(R)\, p(R \to R') = \Psi_T^2(R')\, p(R' \to R) \tag{14}$$

which is fulfilled by the choice $p(R \to R') = 1/N \;\; \min[1, \Psi_T^2(R')/\Psi_T^2(R)]$, with N being the maximum number of possible transitions. It is sufficient to consider only transitions between configurations that are connected by the Hamiltonian, i.e. transitions in which one electron hops to a neighboring site. The standard prescription is then to propose a transition $R \to R'$ with probability $1/N$ and accept it with probability $\min[1, \Psi_T^2(R')/\Psi_T^2(R)]$. This works well for U not too large. For strongly correlated systems, however, the random walk will stay for long times in configurations with a small number of double occupancies $D(R)$, since most of the proposed moves will increase D and hence be rejected with probability $\approx 1 - g^{D(R')-D(R)}$.

There is, however, a way to integrate-out the time the walk stays in a given configuration. To see how, we first observe that for the local energy (12) the ratio of the wavefunctions for all transitions induced by the Hamiltonian have to be calculated. This in turn means that we also know all transition probabilities $p(R \to R')$. We can therefore eliminate any rejection (i.e. accept with probability one) by proposing moves with probabilities

$$\tilde{p}(R \to R') = \frac{p(R \to R')}{\sum_{R'} p(R \to R')} = \frac{p(R \to R')}{1 - p_{\text{stay}}(R)} \, . \tag{15}$$

Checking detailed balance (14) we find that now we are sampling configurations $\tilde{R}_{\rm VMC}$ from the probability distribution function $\Psi_T^2(R)\,(1-p_{\rm stay}(R))$. To compensate for this we assign a weight $w(R)=1/(1-p_{\rm stay}(R))$ to each configuration R. The energy expectation value is then given by

$$E_T \approx \frac{\sum_{\tilde{R}_{\rm VMC}} w(\tilde{R}_{\rm VMC})\, E_{\rm loc}(\tilde{R}_{\rm VMC})}{\sum_{\tilde{R}_{\rm VMC}} w(\tilde{R}_{\rm VMC})}\,. \tag{16}$$

The above method is quite efficient since it ensures that in every Monte Carlo step a new configuration is created. Instead of staying in a configuration where Ψ_T is large, this configuration is weighted with the expectation value of the number of steps the simple Metropolis algorithm would stay there. This is particularly convenient for simulations of systems with strong correlations: Instead of having to do longer and longer runs as U is increased, the above method produces, for a fixed number of Monte Carlo steps, results with comparable error estimates.

Correlated Sampling. So far we have only specified the form of the trial function Ψ_T. The goal of a variational calculation is now to identify the parameters that result in the best trial function. A criterion for a good trial function is e.g. a low variational energy. To find the wavefunction that minimizes the variational energy we could perform independent VMC calculations for a set of different trial functions. It is, however, difficult to compare the energies from these calculations since each VMC result comes with its own statistical errors. This problem can be avoided with correlated sampling [18]. The idea is to use the same random walk for calculating the expectation value of all the different trial functions. This reduces the *relative* errors and hence makes it easier to find the minimum.

Let us assume we have generated a random walk $R_{\rm VMC}$ using Ψ_T as the trial function. Using the same random walk, we can then estimate the energy expectation value (13) for a different trial function $\tilde{\Psi}_T$, by introducing the reweighting factors $\tilde{\Psi}_T^2(R)/\Psi_T^2(R)$:

$$\tilde{E}_T \approx \frac{\sum_{R_{\rm VMC}} \tilde{E}_{\rm loc}(R)\, \tilde{\Psi}_T^2(R)/\Psi_T^2(R)}{\sum_{R_{\rm VMC}} \tilde{\Psi}_T^2(R)/\Psi_T^2(R)}\,. \tag{17}$$

with

$$\begin{aligned}\tilde{E}_{\rm loc}(R) &= \sum_{R'} t\frac{\tilde{\Psi}_T(R')}{\tilde{\Psi}(R)} + U\;D(R) \\ &= \sum_{R'} t\frac{\Psi_T(R')}{\Psi_T(R)}\;\frac{\tilde{\Psi}_T(R')/\Psi_T(R')}{\tilde{\Psi}_T(R)/\Psi_T(R)} + U\;D(R)\end{aligned} \tag{18}$$

We notice that (also in $\tilde{E}_{\rm loc}$) the new trial function $\tilde{\Psi}_T$ appears only in ratios with the old Ψ_T. For trial functions (9) that differ only in the Gutzwiller factor this means that the Slater determinants cancel, leaving only powers $(\tilde{g}/g)^{D(R)}$. Since $D(R)$ is *integer* we can then rearrange the sums in (17) into polynomials in $\tilde{g}/g$. To find the optimal Gutzwiller parameter we then pick a reasonable g, perform a VMC run for $\Psi_T(g)$ during which we also estimate the coefficients for these polynomials. We can then calculate $E(\tilde{g})$ by simply evaluating the ratio of the polynomials. Since there are typically only of the order of some ten non-vanishing coefficients this method is very efficient.

The idea of rewriting the sum over configurations into a polynomial can be easily generalized to trial functions with more correlation factors of the type $r^{c(R)}$, as long as the correlation function $c(R)$ is integer-valued on the space of configurations. As a specific example of how the method works in practice, Fig. 3 shows the result of a correlated sampling run for the trial function (10).

4.2 Fixed-Node Diffusion Monte Carlo

Diffusion Monte Carlo [20] allows us, in principle, to sample the true ground state of a Hamiltonian. The basic idea is to use a projection operator that has the lowest eigenstate as a fixed point. For a lattice problem where the spectrum is bounded $E_n \in [E_0, E_{\max}]$, the projection is given by

$$|\Psi^{(n+1)}\rangle = [1 - \tau(H - E_0)]\,|\Psi^{(n)}\rangle\ ; \quad |\Psi^{(0)}\rangle = |\Psi_T\rangle. \tag{19}$$

If $\tau < 2/(E_{\max} - E_0)$ and $|\Psi_T\rangle$ has a non-vanishing overlap with the ground state, the above iteration converges to $|\Psi_0\rangle$. There is no time-step error involved. Because of the prohibitively large dimension of the many-body Hilbert space, the matrix-vector product in (19) cannot be done exactly. Instead, we rewrite the equation in configuration space

$$\sum |R'\rangle\langle R'|\Psi^{(n+1)}\rangle = \sum_{R,R'} |R'\rangle \underbrace{\langle R'|1 - \tau(H - E_0)|R\rangle}_{=:F(R',R)} \langle R|\Psi^{(n)}\rangle \tag{20}$$

and perform the propagation in a stochastic sense: $\Psi^{(n)}$ is represented by an ensemble of configurations R with weights $w(R)$. The transition matrix element $F(R', R)$ is rewritten as a transition probability $p(R \to R')$ times a normalization factor $m(R', R)$. The iteration (20) is then stochastically performed as follows: For each R we pick a new configuration R' with probability $p(R \to R')$ and multiply its weight by $m(R', R)$. Then the new ensemble of configurations R' with their respective weights represents $\Psi^{(n+1)}$. Importance sampling decisively improves the efficiency of this process by replacing $F(R', R)$ with $G(R', R) = \langle\Psi_T|R'\rangle\, F(R', R)/\langle R|\Psi_T\rangle$, so that transitions from configurations where the trial function is small to configurations with large trial function are enhanced:

$$\sum |R'\rangle\langle\Psi_T|R'\rangle\langle R'|\Psi^{(n+1)}\rangle = \sum_{R,R'} |R'\rangle\, G(R', R)\, \langle\Psi_T|R\rangle\, \langle R|\Psi^{(n)}\rangle. \tag{21}$$

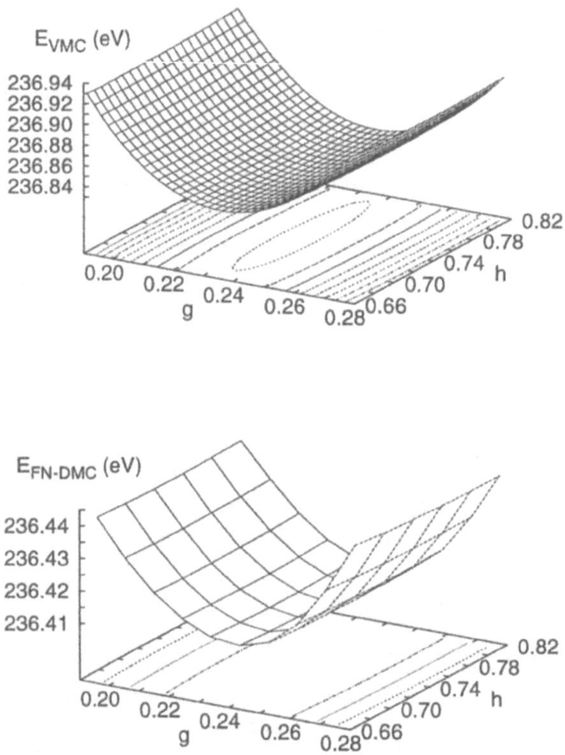

Fig. 3. Correlated sampling for the parameters g and h in the generalized Gutzwiller wavefunction $|\Psi_T\rangle = g^D h^{n_c} |\Phi\rangle$, cf. eqn. (10), in variational (upper plot) and fixed-node diffusion Monte Carlo (lower plot). The plots show the energy as a function of the Gutzwiller parameters g and h, both as surfaces and contours. The calculations were done for an fcc cluster of 64 molecules with 96+96 electrons (half-filled t_{1u}-band), an on-site Hubbard interaction $U = 1.25\,\text{eV}$, and a test charge of $q = 1/4\,\text{e}$

Now the ensemble of configurations represents the product $\Psi_T\,\Psi^{(n)}$. After a large number n of iterations the ground state energy is then given by the mixed estimator

$$E_0 = \frac{\langle\Psi_T|H|\Psi^{(n)}\rangle}{\langle\Psi_T|\Psi^{(n)}\rangle} \approx \frac{\sum_R E_{\text{loc}}(R)\,w(R)}{\sum_R w(R)}. \tag{22}$$

As long as the evolution operator has only non-negative matrix elements $G(R', R)$, all weights $w(R)$ will be positive. If, however, G has negative matrix elements there will be both configurations with positive and negative weight. Their contributions to the estimator (22) tend to cancel so that eventually the statistical error dominates, rendering the simulation useless. This is the infamous sign problem. A straightforward way to get rid of the sign problem is to remove the offending matrix elements from the Hamiltonian, thus defining a new Hamiltonian H_{eff} by

$$\langle R'|H_{\text{eff}}|R\rangle = \begin{cases} 0 & \text{if } G(R',R) < 0 \\ \langle R'|H|R\rangle & \text{else} \end{cases} \tag{23}$$

For each off-diagonal element $\langle R'|H|R\rangle$ that has been removed, a term is added to the diagonal:

$$\langle R|H_{\text{eff}}|R\rangle = \langle R|H|R\rangle + \sum_{R'} \Psi_T(R')\langle R'|H|R\rangle/\Psi_T(R). \tag{24}$$

This is the fixed-node approximation for lattice Hamiltonians [20]. H_{eff} is by construction free of the sign problem and variational, i.e. $E_0^{\text{eff}} \geq E_0$. The equality holds if $\Psi_T(R')/\Psi_T(R) = \Psi_0(R')/\Psi_0(R)$ for all R, R' for which $G(R',R)$ is negative.

Fixed-node diffusion Monte Carlo for a lattice Hamiltonian thus means that we choose a trial function from which we construct an effective Hamiltonian and determine its ground state by diffusion Monte Carlo. Because of the variational property, we want to pick the Ψ_T such that E_0^{eff} is minimized, i.e. we want to optimize the trial function, or, equivalently, the effective Hamiltonian. Also here we can use the concept of correlated sampling. For optimizing the Gutzwiller parameter g we can even exploit the idea of rewriting the correlated sampling sums into polynomials in $\tilde{g}/g$, that we already have introduced in VMC. There is, however, a problem arising from the fact that the weight of a given configuration $R^{(n)}$ in iteration n is given by the product $w(R^{(n)}) = \prod_{i=1}^{n} m(R^{(i)}, R^{(i-1)})$. Each individual normalization factor $m(R',R)$ can be written as a finite polynomial, but the order of the polynomial for $w(R^{(n)})$ increases steadily with the number of iterations. It is therefore not practical to try to calculate the ever increasing number of coefficients for the correlated sampling function $E^{(n)}(\tilde{g})$. But since we still can easily calculate the coefficients for the $m(R',R)$, we may use them to evaluate $E^{(n)}(\tilde{g})$ in each iteration on a set of predefined values $\tilde{g}_i$ of the Gutzwiller parameter. Figure 3 shows an example, again for the more general trial function (10).

Extrapolated Estimator. So far we have only considered estimators for the total energy. For determining the screening, however, we need to know the charge density n_c on site c. It is no problem to calculate the expectation value $n_c(\text{VMC}) = \langle\Psi_T|\hat{n}_c|\Psi_T\rangle$ in variational Monte Carlo. Diffusion Monte Carlo gives, however, only the mixed estimator $n_c(\text{DMC}) = \langle\Psi_T|\hat{n}_c|\Psi_0\rangle$. Since the density operator does not commute with the Hamiltonian, the mixed

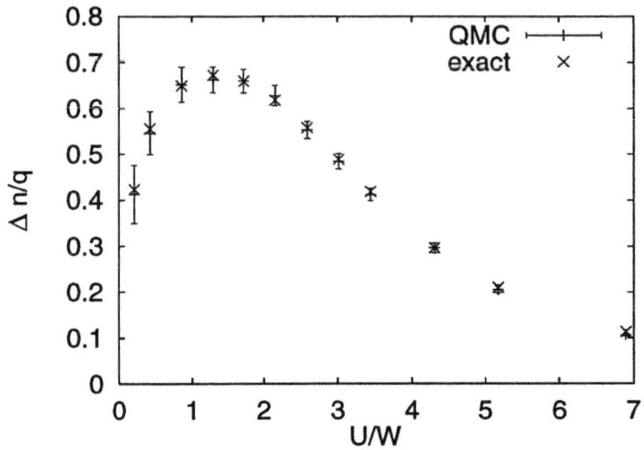

Fig. 4. Screening charge Δn on the site of the test charge ($q = 0.25\,\mathrm{e}$) as a function of U/W, where U is the Coulomb interaction and W is the band width. Exact diagonalization and QMC calculations have been performed for four molecules (12 electrons). The figure shows that the QMC calculations are quite accurate over the whole range of U/W

estimator is different from expectation $n_c(\mathrm{QMC}) = \langle\Psi_0|\hat{n}_c|\Psi_0\rangle$. For a good choice of the trial function the true expectation value can be determined using the extrapolated estimator $n_c(\mathrm{QMC}) \approx 2\,n_c(\mathrm{DMC}) - n_c(\mathrm{VMC})$.

To test the accuracy of this approach to the screening problem, which besides the extrapolated estimator also involves the fixed-node approximation, we have compared the results of the quantum Monte Carlo calculations with the exact results from exact diagonalization for a small system of four molecules (12 electrons). As illustrated by Fig. 4, the QMC calculations for determining the screening charge are very accurate.

Acknowledgment. This work has been supported by the Alexander-von-Humboldt-Stiftung under the Feodor-Lynen-Program and the Max-Planck-Forschungspreis, and by grant No. DEFG02-96ER45439 from the Department of Energy.

References

1. Gunnarsson, O., Koch, E., and Martin, R.M. (1996) Mott transition in degenerate Hubbard models: Application to doped Fullerenes, *Phys. Rev. B* **54**, R11026
2. Gunnarsson, O. (1997) Superconductivity in Fullerides, *Rev. Mod. Phys.* **69**, 575-606
3. Gunnarsson, O., Zwicknagl, G. (1992) Coulomb Pseudopotential, Screening and Superconductivity in C_{60}, *Phys. Rev. Lett.* **69**, 957-960 ; Gunnarsson, O.,

Rainer, D., Zwicknagl, G. (1992) Screened interaction and Coulomb pseudopotential in C_{60}, *Int. J. Mod. Phys.* **6**, 3993-4005
4. Koch, E., Gunnarsson, O., Martin, R.M. (1999) Screening, Coulomb pseudopotential, and superconductivity in alkali-doped Fullerenes, *preprint* , `cond-mat/9902241`
5. Erwin, S. C. (1993) Electronic Structure of the Alkali-Intercalated Fullerides, Endohedral Fullerenes, and Metal-Adsorbed Fullerenes, in W. E. Billups and M. A. Ciufolini (Eds.), *Buckminsterfullerenes*, VCH Publishers, New York, pp. 217-253
6. Löwdin, P. O. (1951) A Note on the Quantum-Mechanical Perturbation Theory, *J. Chem. Phys.* **19**, 1396-1401
7. see e.g. Inversion by Partitioning, in W. H. Press, B. P. Flannery, S. A. Teukolsky, and W. T. Vetterling, *Numerical Recipes in Fortran: The Art of Scientific Computing*, Cambridge University Press, 1992, p. 70
8. Gunnarsson, O., Satpathy, S., Jepsen, O., and Andersen, O. K. (1991) Orientation of C_{60} Clusters in Solids, *Phys. Rev. Lett.* **67**, 3002 ; Satpathy, S., Antropov, V. P., Andersen, O. K., Jepsen, O., Gunnarsson, O., and Liechtenstein, A. I. (1992) Conduction-band structure of alkali-metal-doped C_{60}, *Phys. Rev. B* **46**, 1773
9. Gunnarsson, O., Erwin, S. C., Koch, E., and Martin, R. M. (1998) Role of alkali atoms in A_4C_{60}, *Phys. Rev. B* **57**, 2159
10. Antropov, V. P., Gunnarsson, O., and Jepsen, O. (1992) Coulomb integrals and model Hamiltonians for C_{60}, *Phys. Rev. B* **46**, 13647
11. Brühwiler, P. A., Maxwell, A. J., Nilsson, A., Mårtensson, N., and Gunnarsson, O. (1993) Auger and photoelectron study of the Hubbard U in C_{60}, K_3C_{60}, and K_6C_{60}, *Phys. Rev. B* **48**, 18296
12. Lof, R. W., van Veenendaal, M. A., Koopmans, B., Jonkman, H. T. and Sawatzky, G. A. (1992) Band Gap, Excitons, and Coulomb interaction in solid C_{60}, *Phys. Rev. Lett.* **68**, 3924
13. Mazin, I.I., Rashkeev, S.N., Antropov, V.P., Jepsen, O., Liechtenstein, A.I., and Andersen O.K. (1992) Quantitative theory of superconductivity in doped C_{60}, *Phys. Rev. B* **45**, 5114
14. Hedin, L., and Lundqvist, S. (1969) Effects of Electron-Electron and Electron-Phonon Interactions on the One-Electron States of Solids, in H. Ehrenreich, D. Turnbull and F. Seitz (Eds.), *Solid State Physics*, **23**, Academic Press, New York, pp. 1-181 ; Moroni, S., Ceperley, D.M., and Senatore, G. (1995) Static Response and Local Field Factor of the Electron Gas, *Phys. Rev. Lett.* **75**, 689
15. Ebbesen, T.W., Tsai, J.S., Tanigaki, K., Hiura, H., Shimakawa, Y., Kubo, Y., Hirosawa, I., Mizuki, J. (1992) , *Physica C* **203**, 163; Burk, B., Crespi, V.H., Zettl, A., Cohen, M.L. (1994) , *Phys. Rev. Lett.* **72**, 3706
16. Gutzwiller, M. C. (1963) Effect of correlation on the ferromagnetism of transition metals, *Phys. Rev. Lett.* **10**, 159
17. Horsch, P. and Kaplan, T. A. (1983) Exact and Monte Carlo studies of Gutzwiller's state for the localised-electron limit in one dimension, *J. Phys. C* **16**, L1203; Metropolis, N., Rosenbluth, A. W., Rosenbluth, N. M., Teller, A. H., and Teller, E. (1953) Equation of State Calculations by Fast Computing Machines, *J. Chem. Phys.* **21**, 1087
18. Ceperley, D. M., Chester, G. V., and Kalos, M. H. (1977) Monte Carlo simulation of a many-Fermion study, *Phys. Rev. B* **16**, 3081; Umrigar, C. J., Wilson,

K. G., and Wilkins, J. W. (1988) Optimized Trial Wave Functions for Quantum Monte Carlo Calculations, *Phys. Rev. Lett.* **60**, 1719 Koch, E., Gunnarsson, O., and Martin, R.M (1999) Optimization of Gutzwiller Wavefunctions in Quantum Monte Carlo, *preprint* , `cond-mat/9903070`

19. Trivedi, N. and Ceperley, D. M. (1989) Green-function Monte Carlo study of quantum antiferromagnets, *Phys. Rev. B* **40**, 2737
20. ten Haaf, D. F. B., van Bemmel, H. J. M., van Leeuwen, J. M. J., and van Saarloos, W. (1994) Fixed-node quantum Monte Carlo method for lattice Fermions, *Phys. Rev. Lett.* **72**, 2442; ten Haaf, D. F. B., van Bemmel, H. J. M., van Leeuwen, J. M. J., van Saarloos, W. and Ceperley, D. M. (1995) Proof for an upper bound in fixed-node Monte Carlo for lattice Fermions, *Phys. Rev. B* **51**, 13039 Calandra, M., Sorella, S. (1998) Numerical study of the two-dimensional Heisenberg model using a Green function Monte Carlo technique with a fixed number of walkers, *Phys. Rev. B* **57**, 11446 ; Cosentini, A.C., Capone, M., Guidoni, L., and Bachelet, G.B. (1998) Phase separation in the 2D Hubbard model: a fixed-node quantum Monte Carlo study, *Phys. Rev. B* **58**, R14685

Quantum Simulations of Strongly Correlated Electron Systems

S. Zhang and E. C. Allman

Department of Physics and Department of Applied Science
College of William and Mary, Williamsburg, VA 23187, USA

Abstract. We review recent progress in developing new quantum Monte Carlo methods for simulating strongly correlated electron systems. We then present a summary of our on-going studies of the Hubbard model with these methods, including the *ground-state* constrained path Monte Carlo method and a new *finite-temperature* method. These methods make possible simulations at large system sizes and low temperatures. They eliminate the exponential decay of 'sign' with system size and inverse temperature that is characteristic of the fermion sign problem in standard quantum Monte Carlo. The new methods, which are approximate, yielded accurate results in a variety of benchmarks. On the Hubbard model, our calculations have found no indication of long-range order in electron pairing correlations. An improved band structure in the model, derived from first-principles calculations on $YBa_2Cu_3O_7$, does not appear to enhance pairing.

1 Introduction

One of the most active areas in condensed-matter physics and materials research is the study of strongly correlated quantum-mechanical systems. These include atoms, molecules, clusters, quantum fluids, high-temperature superconductors, heavy-fermion metals, quantum wells and dots, and magnetic materials. The interaction between particles in these systems can lead to such important correlation effects that their properties are crucially modified. High-temperature superconductors are one such example.

In order to understand properties of correlated systems and their theoretical and technological implications, accurate calculations are necessary to correctly treat microscopic correlations. This need is clearly reflected in the daily theoretical demand to solve model systems, which are often intractable analytically.

Monte Carlo simulation methods have shown much promise as a general tool with favorable complexity for studying quantum systems. Rather than explicitly integrating over phase space, they *sample* it. The central limit theorem dictates that the statistical error in a Monte Carlo calculation decays *algebraically* with computer time. With the rapid advent of scalable parallel computers, these methods offer the possibility to systematically study large system sizes in order to extract information about the thermodynamic limit.

For fermion systems, however, Monte Carlo methods suffer from the so-called "sign" problem[1,2]. In these systems, the Pauli exclusion principle

Springer Proceedings in Physics, Volume 85
Computer Simulation Studies in Condensed-Matter Physics XII
Eds.: D.P. Landau, S.P. Lewis and H.-B. Schüttler

requires that the states be anti-symmetric under interchange of two particles. As a consequence, negative signs appear, which cause cancellations among contributions of the Monte Carlo samples of phase space. In fact, as the temperature is lowered or the system size is increased, such cancellation becomes more and more complete. The net signal thus decays *exponentially*. The algebraic scaling is then lost, and the method breaks down. Since most condensed-matter systems consist of fermions, the effect of this is devastating.

The one-band Hubbard model on a square lattice is a microscopic description of a many-electron system. Since the discovery of high-T_c superconductivity, it has been believed that this model contains essential features relevant to the properties of the CuO_2 planes in the cuprate superconductors. The model has been the subject of intense theoretical activity[3], to discover whether it can serve as the paradigm for this novel and important phenomenon.

Quantum Monte Carlo (QMC) has been a promising tool for studying the Hubbard model. The auxiliary-field QMC[4–6] method in particular has been applied extensively and has seen a great deal of success[7,3]. However, due to the sign problem, it has typically been limited to relatively small system sizes, high temperatures, and selected electron fillings. In fact the physically relevant parameter regions (low temperatures and electron filling fractions around 85%) have exhibited the worst sign problem.

Several years ago, Zhang, Carlson, and Gubernatis developed an approximate approach[8,9] to study ground states of correlated electron systems on a lattice. This method, called the constrained path Monte Carlo (CPMC), is free of any sign decay and its computing time scales *algebraically* with system size. In a variety of benchmark calculations, it yielded accurate and encouraging results. The method has seen various applications to study the Hubbard model[10,11], the periodic Anderson model[12], and the three-band Hubbard model[13]. It has also been applied to systems including zeolites[14] and organic superconductors[15].

Very recently, one of us has developed a finite-temperature solution which allows systematic studies of the temperature dependence of correlated-electron models without the exponential growth of computer time with inverse temperature. It retains the field-theoretical formalism[4] which has seen wide applications in condensed-matter, nuclear, and high-energy physics. It allows grand-canonical ensemble simulations. The method is approximate, with a trial propagator to constrain paths in auxiliary-field space. Benchmark calculations[16] show that the algorithm can yield accurate results. This method forms the finite-temperature counterpart to zero-temperature CPMC.

These methods provide opportunities to study correlated electron systems at system sizes and parameter regions that could not be reached before. A first and natural application of these methods is to study the Hubbard model. In what follows, we first describe the new finite-temperature method in Sec. 2.1, and then briefly summarize ground-state CPMC in Sec.2.2. In

Sec. 3, we review their applications to the one-band Hubbard model, focusing on the study of $d_{x^2-y^2}$-wave electron pairing correlations. We then make some concluding remarks in Sec. 4.

2 New Quantum Monte Carlo Methods

2.1 Finite-Temperature Method

The new finite-temperature method[16] is free of any decay of the average sign, thus making possible calculations at lower temperatures and larger system sizes with favorable scaling. The method retains many of the advantages of the standard formalism. It allows grand-canonical ensemble simulations. With the second-quantization representation, it also automatically accounts for particle permutations. Both diagonal and off-diagonal expectations are easily computed, as well as imaginary-time correlation functions.

Below we first recapitulate the formalism of Blankenbecler, Scalapino, and Sugar (BSS)[4,6], in a way that will facilitate our introduction of the new method. We then discuss the new method and present some benchmark data. Further details can be found in Ref. [16].

The Standard BSS Algorithm Given a Hamiltonian H, the expectation value of a physical observable O is:

$$\langle O \rangle = \frac{\mathrm{Tr}(Oe^{-\beta H})}{\mathrm{Tr}(e^{-\beta H})}, \tag{1}$$

where $\beta = 1/kT$ is the inverse temperature. The chemical potential term is implicit in the Hamiltonian H. We can apply the Trotter break-up and write the denominator, namely the partition function Z, as

$$Z \equiv \mathrm{Tr}(e^{-\beta H}) = \mathrm{Tr}[\underbrace{e^{-\Delta\tau H} \cdots e^{-\Delta\tau H} e^{-\Delta\tau H}}_{L}], \tag{2}$$

where $\Delta\tau = \beta/L$ is the time-step and L is the number of "time slices."

The next step is to write the short-time or high-temperature propagator $e^{-\Delta\tau H}$ in terms of one-electron (non-interacting) propagators. This is possible for most Hamiltonians or Euclidean actions of interest. For example, the Hubbard-Stratanovic transformation[17] can be applied for a Hamiltonian H which contains one- and two-body terms, denoted by K and V, respectively. This transformation replaces the two-body term $e^{-\Delta\tau V}$ by one-body interactions with a set of random external fields. Combining the result with the one-body term $e^{-\Delta\tau K}$, we can write

$$e^{-\Delta\tau H} \simeq \sum_{\mathbf{x}} B(\mathbf{x}), \tag{3}$$

where $\mathbf{x}$ denotes the random external auxiliary fields and $B(\mathbf{x})$ is a *single-particle operator*. With no loss of generality we have written the integration over $\mathbf{x}$ as a discrete sum. We have also suppressed spin indices. The approximation in Eq. (3) is from the Trotter error, which is of $\mathcal{O}(\Delta\tau^2)$ or less. The essence of the HS transformation in Eq. (3) is the conversion of an interacting system into many non-interacting systems living in fluctuating external auxiliary-fields, and the summation over all such auxiliary-field configurations recovers the many-body interactions.

Substituting Eq. (3) into Eq. (2) leads to

$$Z = \sum_X \mathrm{Tr}[B(\mathbf{x}_L)\cdots B(\mathbf{x}_2)B(\mathbf{x}_1)], \tag{4}$$

where $X \equiv \{\mathbf{x}_1, \mathbf{x}_2, \cdots, \mathbf{x}_L\}$ denotes a complete *path* in auxiliary-field space. Since the Trotter error in Eq. (3) is controllable (by, e.g., extrapolation with different values of $\Delta\tau$), we have used the equal sign above. The trace over fermion degrees of freedom is then performed analytically[4,18], which yields

$$Z = \sum_X \det[I + B(\mathbf{x}_L)\cdots B(\mathbf{x}_2)B(\mathbf{x}_1)]. \tag{5}$$

If the size of the single-particle basis (*e.g.*, number of spatial lattice sites) is N, the single-particle propagator $B(\mathbf{x}_l)$ is an $N \times N$ matrix and I is the corresponding unit matrix. Note that the the trace in Eq. (4), which is over different numbers of particles and initial states, is now replaced by the fermion determinant. The fermion determinant, which we will denote by $D(X)$, can be readily computed for each path X. The sum over all paths is evaluated by standard Monte Carlo techniques, for example, the Metropolis algorithm.

The sign problem occurs because $D(X)$ is not always positive. In practice, the MC samples of X are drawn from the probability distribution defined by $|D(X)|$. As β increases, $D(X)$ approaches an antisymmetric function and its average sign,

$$\langle s\rangle = \sum_X D(X)/\sum_X |D(X)|, \tag{6}$$

vanishes exponentially as β increases. The variance in the MC estimate of Eq. (1) thus diverges, and the sign problem occurs.

The New Finite-T Method The implicit nature of the formalism above poses difficulties in understanding how to control the sign problem. Paths in the grand-canonical formulation, contrary to ground-state projector QMC, do not originate or end at a single explicit point in Slater determinant space. Instead they involve many points, due to the trace over states with varying number of particles. Indeed the points do not even have the same "dimension." A straightforward application of the ideas from ground-state CPMC[8,9] would require separate constraining conditions for paths with different end points.

Even if such constraints are found, for instance in the form of a density matrix as in restricted path-integral[19] Monte Carlo, their implementation appear difficult due to the analytical evaluation of the trace over the end points. More specifically, we must address the following issues:

- What is the origin of the vanishing $\langle s \rangle$ and how precisely does it occur in the sampling process?
- How do we introduce physically motivated constraints that will make $\langle s \rangle$ non-vanishing, given that the path-integral picture is *implicit* in Eq. (5)?
- How can we implement any constraints, given that multiple states are involved in the trace and that they have variable dimensions because of the canonical ensemble formulation?
- Because the observable requires the entire path to compute (as opposed to ground-state calculations in which only the "instantaneous wave function" after convergence is needed), the constraints on the path must be non-local in imaginary time. How can we implement such non-local conditions efficiently?

The new finite-T algorithm answers all of the above. To describe the method, we introduce $\mathcal{B}$ to denote the propagator in Eq. (3), i.e., $\mathcal{B} \equiv e^{-\Delta\tau H}$, and

$$\mathcal{P}_l(\{\mathbf{x}_1, \mathbf{x}_2, \cdots, \mathbf{x}_l\}) \equiv \mathrm{Tr}[\underbrace{\mathcal{B}\mathcal{B}\cdots\mathcal{B}}_{L-l}\, B(\mathbf{x}_l)\cdots B(\mathbf{x}_2)B(\mathbf{x}_1)] \tag{7}$$

to denote the overall contribution in the partition function of a *partial path* of length l: $\{\mathbf{x}_1, \mathbf{x}_2, \cdots, \mathbf{x}_l\}$. It can be shown[16] that constraining the path X with the following condition

$$\begin{aligned} \mathcal{P}_1(\{\mathbf{x}_1\}) &> 0 \\ \mathcal{P}_2(\{\mathbf{x}_1, \mathbf{x}_2\}) &> 0 \\ &\vdots \\ \mathcal{P}_L(\{\mathbf{x}_1, \mathbf{x}_2, \cdots, \mathbf{x}_L\}) &> 0 \end{aligned} \tag{8}$$

leads to identical result of Z in Eq. (4) as the original, unconstrained sum. That is, the partition function can be obtained with a constrained sum over only a subset of all possible paths X. Further, each path in this subset leads to a final contribution in Z given by: $P_L(X) \equiv D(X) > 0$, due to the last condition in (8). Hence the average sign in $\langle s \rangle$ would remain 1 and the sign problem would be solved!

The problem is of course we do not know $\mathcal{B}$. So, while we have clearly revealed the origin of the sign problem with (8), we do not have a practical solution yet. The next step is to approximate $\mathcal{B}$ by a *trial propagator* B_T. The constraints now yield approximate results, which become exact if B_T is exact. If B_T is in the form of a single-particle propagator or a linear combination of single-particle propagators, the trace in Eq. (7) can be analytically evaluated by use of the same identity[18] that produced Eq. (5). The constraints in (8) can now be written as:

$$\mathcal{P}_l^T(\{\mathbf{x}_1, \mathbf{x}_2, \cdots, \mathbf{x}_l\}) = \det[I + (\prod_{m=1}^{L-l} B_T)\, B(\mathbf{x}_l) \cdots B(\mathbf{x}_1)] > 0 \qquad (9)$$

for each l on $1 \le l \le L$.

An efficient algorithm[16] can be constructed which naturally and directly incorporates the constraints into the sampling process. Each path X is generated by a random walk of L steps, with $\mathbf{x}_l$ sampled in the lth step and conditional on the previous $l-1$ fields. The sampling of each $\mathbf{x}_l$ is done both with the constraint and with importance sampling[16,20]. The latter allows $\mathbf{x}_l$ to be selected according to the best estimate of potential contribution in Z.

2.2 $T = 0\,\mathrm{K}$ Constrained-Path Monte Carlo (CPMC)

The algorithm we have described provides the *finite-temperature* counterpart of the *ground-state* constrained path Monte Carlo (CPMC) method[8] developed several years ago. The latter has been applied to the one-band Hubbard model, whose results we will review and discuss briefly in the next section, the three-band Hubbard mode, Hubbard ladders, the periodic Anderson model, and models for systems including zeolites and organic superconductors. The method eliminated the sign decay in $T = 0\,\mathrm{K}$ auxiliary-field calculations by constraining paths in Slater determinant space with a trial ground-state wave function $|\psi_T\rangle$ [8]. Here we will briefly summarize the $T = 0\,\mathrm{K}$ CPMC method, and refer the reader to Refs.[8,9] for details.

The ground-state CPMC algorithm has two main components: The *first* is to formulate the projection of the ground state as open-ended random walks with importance sampling, as in Green's function Monte Carlo[20], although the random walks take place in a space of Slater determinants. The *second* is to constrain the paths of the random walks so that any Slater determinant generated maintains a positive overlap with a known trial wave function $|\psi_T\rangle$. The first is an alternative way to do projector QMC which in many cases can be more efficient than the latter, because of the use of importance sampling. Like projector QMC, it is exact, but suffers from the sign problem. The second component is the constrained path approximation, which eliminates the sign decay, but introduces a systematic error in the algorithm. These two components are independent of each other, and can be used separately. Their combination is what we are calling the ground-state CPMC algorithm.

CPMC is free of any "sign decay", which is the signature of the fermion sign problem. The required computation time therefore scales algebraically with system size. It is approximate, with results dependent on the trial wave function that is used as constraint. Similar to the fixed-node approximation [21,22] in configuration space, the process of CPMC solves the Schroedinger equation under some boundary condition defined by the trial wave function. But, differently from fixed-node, this boundary condition is not limited to real space, and can be a more "global" one.

CPMC also retains (at least to a large degree) an important advantage of the auxiliary-field QMC formalism compared to continuum methods, namely the ability to compute easily many expectation values, including off-diagonal ones such as electron pairing correlations. This is an important point since in many applications, particularly those in the study of strongly correlated models, it is crucial to have information on properties beyond the total energy.

In a variety of benchmark calculations, on the Hubbard model in 1D and 2D, CPMC has produced very accurate estimates of the energy as well as other ground-state observables. Practically, the systematic error of CPMC, even for most small (6×6) systems, is smaller than the statistical error of the standard lattice algorithms. The latter increases rapidly with system size. (At 12×12, to achieve the same statistical error as CPMC, PQMC would have to run at least four orders of magnitude longer.[23,8])

As discussed in Sec. 2.1, the concept of a constraining wave function $|\psi_T\rangle$ was difficult to generalize to finite temperatures. On the other hand, the formalism we developed in the new finite-T method, which is based on the propagator $\mathcal{B}$ rather than states, does allow a more a unified view of the two algorithms. The constraining $|\psi_T\rangle$ in $T = 0\,\mathrm{K}$ CPMC can be understood as the outcome of B_T in the finite-T method operating on an initial state.

3 Is the Hubbard Model the Right One for High-T_c?

3.1 The Hubbard Model

The one-band Hubbard model on a square lattice is a simple microscopic description of a many-electron system which includes both band-structure and electron interactions. Since the discovery of high-T_c superconductivity, the model has been believed[24,3] to contain the essential features relevant to the properties of the CuO_2 planes in the cuprate superconductors. Viewed as the basis for any further progress on understanding the mechanism for high-T_c[24], the model has been the subject of an unprecedented level of theoretical activity[3,25]. The effort has lead to much insight and understanding. It also resulted in a good deal of available data, which makes the model a good testing ground for new methods.

The basic Hamiltonian is

$$H = -t \sum_{\langle ij \rangle \sigma} (c^\dagger_{i\sigma} c_{j\sigma} + c^\dagger_{j\sigma} c_{i\sigma}) + U \sum_i n_{i\uparrow} n_{i\downarrow} + \mu \sum_i (n_{i\uparrow} + n_{i\downarrow}), \tag{10}$$

where t is the electron hopping matrix element between near-neighbors and $c^\dagger_{i\sigma}$ and $c_{i\sigma}$ are electron creation and destruction operators, respectively, of spin σ on site i. The notation $\langle\ \rangle$ indicates near-neighbors. The on-site

Coulomb repulsion is $U > 0$, and $n_{i\sigma} = c_{i\sigma}^{\dagger} c_{i\sigma}$ is the electron number operator. We will denote the number of lattice sites by N, and the linear dimension by L. The numbers of electrons with spin $\sigma = \uparrow, \downarrow$, which we denote by $N_{\uparrow}$ and $N_{\downarrow}$. In *ground-state calculations* they will be fixed, the last term involving μ does not appear in the Hamiltonian, and we work in the canonical ensemble. In *finite-T calculations*, the number of electrons is controlled by the chemical potential μ, and we work in the grand canonical ensemble.

Below we list a few results on this model which are directly relevant to the ones we will discuss. Clearly there has been much more work than we could even begin to describe here[3].

1. Certain normal-state properties of the cuprate superconductors can be understood within the framework of this model. Experimental results are consistent with the notion that the dominant quasi-particle scattering mechanism in the cuprates involves spin-fluctuations — the model provides a description of these spin-fluctuation processes[26,3]. Monte Carlo calculations at half-filling $\langle n \rangle = 1$ (no sign problem due to particle-hole symmetry) established the presence of long-range anti-ferromagnetic order[6].
2. Various calculations[25] have suggested the possible appearance of superconductivity in this model upon doping ($\langle n \rangle < 1$). Further, there is evidence that, *if* superconductivity indeed occurs, the superconducting state will have $d_{x^2-y^2}$ symmetry. The calculations range from various approximate analytic approaches to exact diagonalizations of small systems (4×4) to quantum Monte Carlo (QMC) simulations.
3. The QMC calculations in 2 found an effective electron-electron interaction which is attractive in the singlet $d_{x^2-y^2}$ channel. However, they did not find evidence for a condensate of pairs, namely long-range pairing correlations or finite superfluid density. The calculations were severely limited[23,27] by the sign problem, e.g., to temperatures of about $T \sim t/6$.

3.2 Finite-T Benchmark Calculations

In Fig. 1, we show results on total and kinetic energies per site, for a simple case of the 4×4 system. The chemical potential is adjusted so that the average filling fraction corresponds to 5 $\uparrow$ and 5 $\downarrow$ electrons. At high temperatures, the new algorithm reproduces the results from the standard method of BSS, while at low temperatures it does not suffer from any decay of sign and smoothly extends to ground-state result from exact diagonalization. Results from the new method are shown up to $\beta = 16$ ($T = 0.0625$).

In Fig. 2, we show results for a 4×4, $U = 4$ system where the sign problem is the most severe. In the top figure $\langle s \rangle$ in BSS is shown for several values of T, directly measured in calculations at high temperatures and projected from the exponential decay rate[2] at low temperatures ($\beta = 12$ and 20). The sign problem limits the range of temperatures where accurate calculations can be

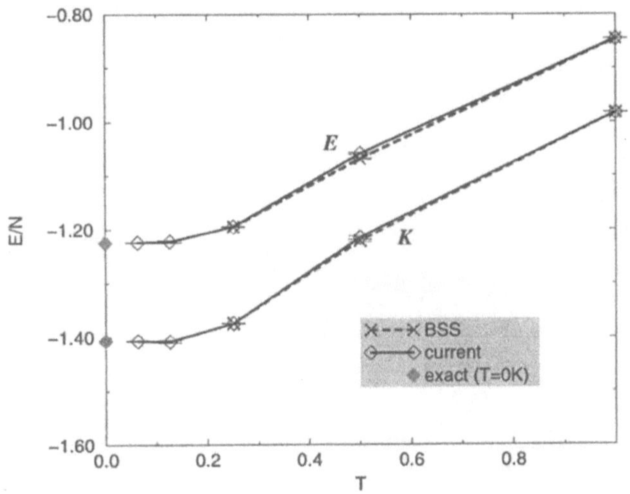

Fig. 1. Benchmark of new finite temperature method on a small system (4×4) with $U = 4$ and density $\langle n \rangle = 0.875$. Shown are total and kinetic energies per site at different temperatures T. To gauge the accuracy, we show at $T = 0.5$ ($\beta = 2$) results by BSS and the new algorithm at finite $\Delta\tau$, i.e., prior to extrapolation to $\Delta\tau \to 0$. (Because of the particular ways of doing the Trotter break-up in the two methods, their finite-$\Delta\tau$ errors are in opposite directions.) The diamond at $T = 0$ is from exact diagonalization

carried out with the standard algorithm. The system hence presents a highly challenging test case for the current algorithm.

At high T, our algorithm gives results in excellent agreement with BSS results[28], which are exact. At low T, it reaches convergence and leads to results consistent with those from ground-state CPMC and in good agreement with those from $T = 0\,\mathrm{K}$ exact diagonalization[29]. To illustrate the accuracy of the results and the extent of the agreement between the new method and BSS, we show at $\beta = 4$ the computed E/N from BSS with a finite $\Delta\tau$ of 0.1, which is a typical choice in most calculations, in addition to the results obtained after performing the extrapolation to $\Delta\tau \to 0$. The agreement in the bottom figure is especially reassuring, where the new method, with a wrong B_T, correctly and accurately predicts the evolution of the near-neighbor density-density correlation between opposite spins as T is lowered.

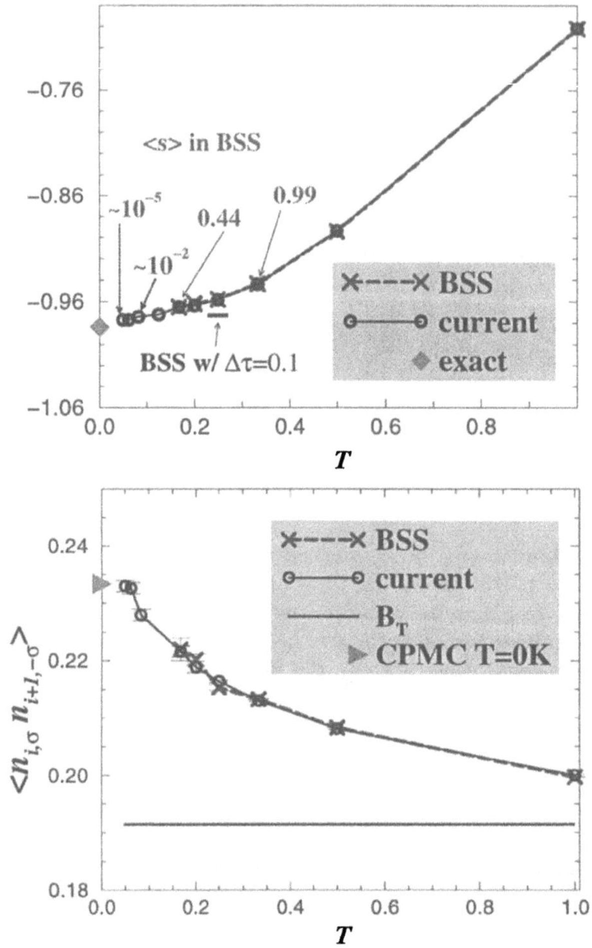

Fig. 2. Benchmark of new finite temperature method on a small system (4×4) with $U = 4$ and density $\langle n \rangle = 0.875$. The sign problem in this case is severe in the standard BSS algorithm. Numbers in the top figure indicate the average sign $\langle s \rangle$ in BSS, directly measured at $\beta = 3$ and 6, and projected from the exponential decay at $\beta = 12$ and 20. *The top figure* shows energies per site at different temperatures T. The bar at $\beta = 4$ indicate the BSS result at $\Delta\tau = 0.1$, i.e., prior to extrapolation to $\Delta\tau \to 0$. The diamond at $T = 0$ is from exact diagonalization. Statistical errors are smaller than symbol size, and are not shown. *The bottom figure* shows the density-density correlation function between near-neighbor sites. The algorithm accurately predicts the development of strong antiferromagnetic correlation as T decreases, despite the use of a constraining propagator B_T which by itself gives incorrect physics (flat line). At low T, the results converge to that of ground-state CPMC

3.3 Simulation Results on Electron Pairing Correlations

The pairing correlation function we computed is defined as

$$P_\alpha(\mathbf{l}) = \langle \Delta_\alpha^\dagger(\mathbf{l}) \Delta_\alpha(\mathbf{0}) \rangle, \tag{11}$$

where α indicates the nature of pairing. The pair-field operator at site $\mathbf{l}$ is $\Delta_\alpha(\mathbf{l}) = \sum_\delta f_\alpha(\delta)[c_{\mathbf{l}\uparrow} c_{\mathbf{l}+\delta\downarrow} - c_{\mathbf{l}\downarrow} c_{\mathbf{l}+\delta\uparrow}]$, where δ is $(\pm 1, 0)$ and $(0, \pm 1)$. For the extended s-wave ($s^\star$-), $f_{s^\star}(\delta) = 1$. For d-wave, $f_d(\delta)$ is 1 when $\delta = (\pm 1, 0)$ and -1 otherwise. Here we will focus on the d-wave electron pairing correlation.

Because both the $T = 0\,\mathrm{K}$ and the finite-T methods have an uncontrolled approximation and because the electron pairing correlations are extremely difficult quantities to characterize, we have carried out benchmark calculations[10,8,16,9] and many self-consistency checks. Due to the limited availability of data for benchmark, the latter becomes even more important. In Fig. 3, we show such an example[9] in ground-state calculations, for a fairly difficult case of 8×8 with $27 \uparrow\ 27 \downarrow$ electrons.

Fig. 4 shows the $d_{x^2-y^2}$-wave pairing correlations for a 12×12 lattice at $U = 2$ and 4, for distances $|l| \geq 3$. Here, the electron filling is 0.85, which corresponds to a closed shell case with $N_\uparrow = N_\downarrow = 61$. We used the free-electron wave function as $|\Psi_T\rangle$ in CPMC. We see that the computed $d_{x^2-y^2}$-wave pairing correlations at both values of U are bounded from above by the $U = 0$ result. Consistent with this observation, we see that $P_d(l)$ decreases as U increases from 2 to 4. We mention that, as U reaches 8, the error bars are large, but $P_d(l)$ oscillates around zero and shows no indication of enhancement[10].

In Fig. 5, we increase the lattice size to 16×16. Here, for a closed shell case with the same electron filling of 0.85 ($N_\uparrow = N_\downarrow = 109$), we show the CPMC results for $U = 2$ and 4. We see that the same trend as in Fig. 4 persists for $P_d(l)$. As the interaction strength is increased to 4, the long-range correlations vanish. Although the error bars (1 standard deviation throughout the paper) are large, they are (almost) good enough to show the irregular oscillations of $P_d(l)$ around zero.

We also note that the behavior of $P_d(\mathbf{l})$ in Fig. 5 at short distances is just the opposite of that at large distances, i.e., $P_d(\mathbf{l})$ increases as U is increased. Thus, due to the large discrepancy in magnitude between short and long distances, the integrated value of $P_d(\mathbf{l})$ (i.e., the structure factor at $(0,0)$) would not give the correct trend even at system sizes as large as these.

We have also studied[11] a modified Hubbard model with improved band structure. The model now contains additional next-nearestt' and third-nearest neighbor hopping t'' terms. The values for these additional terms were chosen based on first-principles calculations on the material $YBa_2Cu_3O_7$[30]. In Fig. 6, we show results for $T = 0\,\mathrm{K}$ CPMC calculations on an 8×8 system with 84% filling. We see that here the error bars $U = 8$ are small and there is no indication of enhancement. This preliminary data would suggest that the improved band structure does not alter our picture above on d-wave pairing.

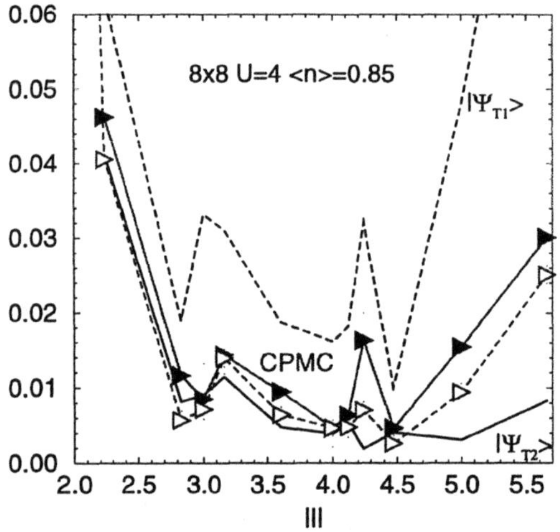

Fig. 3. Sensitivity of CPMC results to the choice of trial wave function $|\Psi_T\rangle$. Shown is the near-neighbor d-wave pairing correlation as a function of pair separation $|\mathbf{l}|$. The short distance portion is omitted. This system corresponds to a difficult open-shell case. The two trial wave functions are both unrestricted Hartree-Fock wave functions. The first, $|\psi_{T1}\rangle$, was obtained with a very small interaction strength U, resembling a free-electron wave function, while $|\psi_{T2}\rangle$ was generated with the true U of 4. The CPMC pairng correlations, with $|\psi_{T1}\rangle$ and $|\psi_{T2}\rangle$ as trial wave functions respectively, are shown (curves with symbols (red)), with the dashed one from $|\psi_{T1}\rangle$. The mean-field results that $|\psi_{T1}\rangle$ and $|\psi_{T2}\rangle$ yield are also shown (no symbols). We see that the two sets of CPMC results, although not in exact agreement, over-correct the mean-field results and show consistency. The statistical error bars on the CPMC results are indicated by small dots, the maximum being several times the symbol size

We are presently doing calculations with the new finite-T method. Using local workstations, we have obtained results of excellent statistical accuracy on systems of size up to 8×8. Fig. 7 shows one set of results on d-wave pairing correlations at the most difficult filling fraction (i.e., with the most severe sign problem) of 0.82.

These results suggest that the near-neighbor $d_{x^2-y^2}$-wave electron pairing correlation does not show long-range order. No indication of enhancement is found over the non-interacting system. Indeed, the behavior of the pairing correlation appears similar across a fairly broad range of electron fillings, including the half-filled case, which exhibits pairing correlations of similar magnitude[10]. Preliminary calculations with an improved band structure show no indication of enhancement. Preliminary finite-T results seem to be consistent with ground-state results.

The new methods enabled calculations on large systems and low temperatures (including $T = 0\,\mathrm{K}$). This latter parameter is particularly significant,

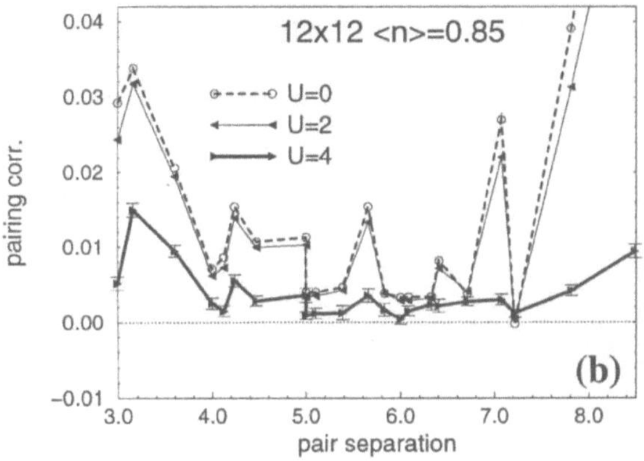

Fig. 4. Ground-state CPMC results on $d_{x^2-y^2}$-wave electron pairing[10]. Figure shows dependence of pairing on U, for a 12×12 system. Pairing decays as U is increased

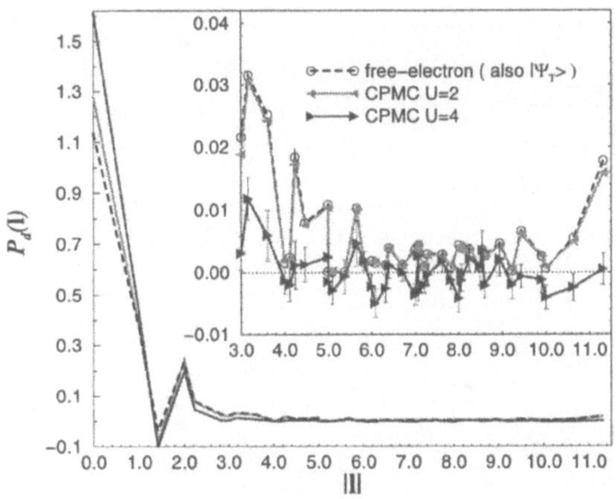

Fig. 5. Ground-state CPMC results on $d_{x^2-y^2}$-wave electron pairing[10]. Figure shows dependence of pairing on U, for a 16×16 system. The main graph is the pairing correlation function vs. pair separation for the entire range of $|\mathbf{l}|$. The inset shows only the long-range behavior starting from $|\mathbf{l}| = 3$, similar to Fig. 4. Pairing shows same trend as that of 12×12, but now the long-range correlation vanishes at $U = 4$

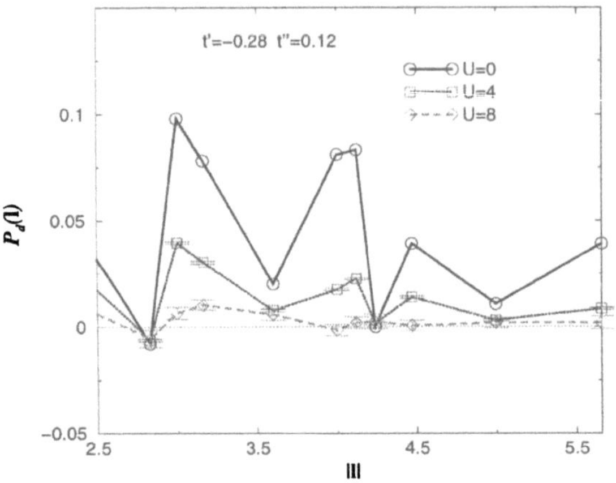

Fig. 6. Ground-state CPMC results on $d_{x^2-y^2}$-wave electron pairing for the t-t'-t'' model. Figure shows dependence of pairing on U, for an 8×8 system, with 27 ↑ and 27 ↓ electrons. Again pairing decays as U is increased

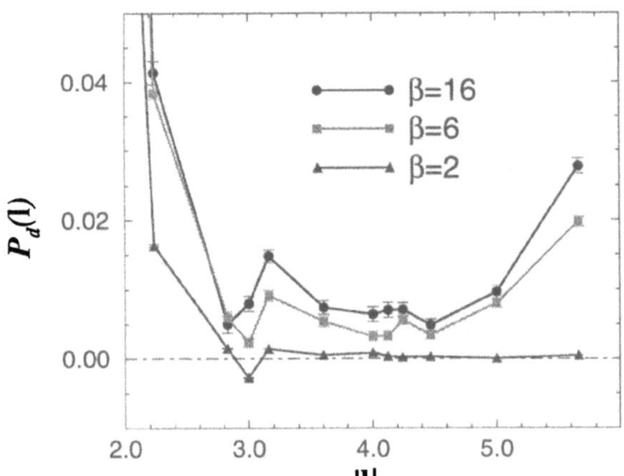

Fig. 7. Finite-temperature results on $d_{x^2-y^2}$-wave electron pairing on 8×8 system with $U = 4$ at an average filling of 0.82, chosen because it corresponds to the worst sign problem in BSS[6]. Figure shows dependence of pairing on T. Pairing increases as T is decreased. However, the non-interacting system also shows the same behavior. In fact the $P_d(|l|)$ in the latter is larger than the corresponding interacting results, which is consistent with ground-state CPMC results shown above

since previous results (Sec. 3.1), and an estimate of $T_c \sim 0.02t$ (from analytic approximations), had led to a strong suspicion that the absence of signals for long-range pairing in QMC calculations was simply due to the inability to reach low enough temperatures[3,25,31].

We emphasize again that, due to the approximate nature of the methods, the possibility can not be completely ruled out that the results were biased such that the pairing correlation was suppressed. However, the many benchmark calculations, and particularly the variety of self-consistency checks, indicate strongly that the results are accurate and robust.

4 Concluding Remarks

The new finite-temperature method and its $T = 0\,\mathrm{K}$ counterpart, the ground-state CPMC, offer exciting opportunities for studying correlated electron systems. A variety of calculations are now possible — while we must be exceedingly careful to understand the approximations involved, no longer are we limited to small system sizes, or high temperatures, or certain regions of the parameter space.

On the Hubbard model, our calculations have found no indication of long-range order in electron pairing correlations. An improved band structure in the model does not appear to enhance pairing. While much more work is needed to check these preliminary results and clarify the role of the approximation, the results to date, unfortunately, raise the question again on whether the model could indeed serve as the basis for understanding high-T_c.

Acknowledgements

We thank J. Carlson and J. E. Gubernatis for collaborations and discussions. We gratefully acknowledge support from the US National Science Foundation under a CAREER Award (DMR-9734041) and from Research Corporation under a Research Innovation Award. Part of the calculations were performed on computers at the NCSA (UIUC).

References

1. K. E. Schmidt and M. H. Kalos, in *Applications of the Monte Carlo Method in Statistical Physics*, edited by K. Binder (Springer Verlag, Heidelberg, 1984).
2. E. Y. Loh Jr., J. E. Gubernatis, R. T. Scalettar, S. R. White, D. J. Scalapino, and R. L. Sugar, Phys. Rev. B **41**, 9301 (1990).
3. See, e.g., E. Dagotto, Rev. Mod. Phys. **66**, 763 (1994) and references therein.
4. R. Blankenbecler, D. J. Scalapino, and R. L. Sugar, Phys. Rev. D **24**, 2278 (1981).
5. G. Sugiyama and S.E. Koonin, Ann. Phys. (NY) **168**, 1 (1986).

6. S. R. White, D. J. Scalapino, R. L. Sugar, E. Y. Loh Jr., J. E. Gubernatis, R. T. Scalettar,, Phys. Rev. B **40**, 506 (1989).
7. D. J. Scalapino, in *High Temperature Superconducting Proceedings*, ed. by K. S. Bedell, D. Coffey, D. E. Meltzer, D. Pines, and J. R. Schrieffer (Addison Wesley, 1990), p314.
8. Shiwei Zhang, J. Carlson, and J. E. Gubernatis, Phys. Rev. Lett. **74**, 3652 (1995); Phys. Rev. B **55**, 7464 (1997).
9. Shiwei Zhang, in *Quantum Monte Carlo Methods in Physics and Chemistry*, Ed. M. P. Nightingale and C. J. Umrigar, NATO ASI Series (Kluwer Academic Publishers, 1999).
10. Shiwei Zhang, J. Carlson, and J. E. Gubernatis, Phys. Rev. Lett. **78**, 4486 (1997).
11. E. C. Allman and Shiwei Zhang, to be published.
12. J. Bonca and J. E. Gubernatis, Phys. Rev. B **58**, 6992 (1998)
13. M. Guerrero, J.E. Gubernatis, and Shiwei Zhang, Phys. Rev. B **57**, 11980 (1998)
14. Otto F. Sankey, A. A. Demkov, and T. Lenosky, Phys. Rev. B **57**, 15129 (1998).
15. S. Mazumdar, S. Ramasesha, R.T. Clay, D.K. Campbell, Phys. Rev. Lett. **82**, 1522 (1999)
16. Shiwei Zhang, preprint (submitted to Phys. Rev. Lett.).
17. J. Hubbard, Phys. Rev. Lett. **3**, 77 (1959).
18. J. E. Hirsch, Phys. Rev. B **31**, 4403 (1985).
19. D. M. Ceperley, Phys. Rev. Lett. **69**, 331 (1992); in *Simulation in Condensed Matter Physics and Chemistry*, Ed. by K. Binder and G. Ciccotti (1996).
20. M. H. Kalos, D. Levesque, and L. Verlet, Phys. Rev. **A9**, 2178 (1974); D. M. Ceperley and M. H. Kalos, in *Monte Carlo Methods in Statistical Physics*, edited by K. Binder (Springer-Verlag, Heidelberg, 1979).
21. J. B. Anderson, J. Chem. Phys. **63**, 1499 (1975); **65**, 4122 (1976).
22. J. W. Moskowitz, K. E. Schmidt, M. A. Lee, and M. H. Kalos, J. Chem. Phys. **77**, 349 (1982); P. J. Reynolds, D. M. Ceperley, B. J. Alder, and W. A. Lester, J. Chem. Phys. **77**, 5593 (1982).
23. N. Furukawa and M. Imada, J. Phys. Soc. Jpn. **61**, 3331 (1992).
24. See, e.g., P. W. Anderson and J. R. Schrieffer, "The give-and-take between two solid-state therists offers insight ... developing a new theroy", Physics Today June, 55 (1991).
25. See, for example, D. J. Scalapino, Phys. Reports **250**, 329 (1995), and references therein.
26. For example, see A. P. Kampf, Phys. Reports **249**, 219 (1994).
27. A. Moreo and D. J. Scalapino, Phys. Rev. B **43**, 8211 (1991); A. Moreo, Phys. Rev. B **45**, 5059 (1992).
28. R. T. Scalettar, private communication.
29. A. Parola, S. Sorella, S. Baroni, R. Car, M. Parrinello, and E. Tosatti, Physica C **162-164**, 771 (1989); A. Parola, S. Sorella, M. Parrinello, and E. Tosatti, Phys. Rev. B **43**, 6190 (1991); G. Fano, F. Ortolani, and A. Parola, Phys. Rev. B **42**, 6877 (1990).
30. O. K. Andersen, A. I. Liechtenstein, O. Jepsen, and F. Paulsen, J. Phys. Chem. Solids **56**, 1573 (1995).
31. D. J. Scalapino, in *Perspectives in Many-Body Physics*, edited by R. A. Broglia, J. R. Schrieffer, and P. F. Bortignon (North-Holland, Amsterdam, 1994), pg. 95.

Competing Effects in Perovskite Manganites: Ferromagnetism vs. Phase Separation

S. Yunoki, A. Moreo, and E. Dagotto

National High Magnetic Filed Laboratory, Florida State University, Tallahassee FL 32310, USA

Abstract. This paper is an informal and brief review of the present status of the computational studies of models for manganites that recently reported the existence of phase separation at low temperatures (S. Yunoki et al., Phys. Rev. Lett. **80**, 845 (1998); **81**, 5612 (1998); A. Moreo et al., cond-mat/9901057 (to appear in Science)). Results for both the one-orbital and two-orbital models, the latter with Jahn-Teller phonons, are included. The main finding is that the transition between the insulating state of the hole-undoped limit and the ferromagnetic metallic state at finite hole-density occurs through a mixed-phase process. When long-range Coulomb interactions are included, a microscopically charge inhomogeneous state should be stabilized. These phase separation tendencies, also present at low electronic densities, influence the properties of the ferromagnetic region by increasing charge fluctuations. A large list of experimental results are compatible with the existence of charge-inhomogeneities in the manganites, which may be an important ingredient for the explanation of the CMR effect in these compounds.

1 Introduction

Colossal magnetoresistance in manganese oxides with a perovskite structure such as $R_{1-x}A_xMnO_3$ (where R = La, Pr, Nd; A = Sr, Ca, Pb) has been attracting considerable attention [1] due to its fundamental importance in the field of condensed matter physics as well as its potential technological applications. A variety of experiments have revealed that oxide manganites have a rich phase diagram [2] with regions of antiferromagnetic (AF) and ferromagnetic (FM) order, as well as charge ordering, and a peculiar insulating state above the FM critical temperature T_C.

The appearance of ferromagnetism at low temperatures can be explained qualitatively using the double exchange (DE) mechanism [3–5]. However, the DE framework is incomplete to describe the entire phase diagram observed experimentally. For instance, the electron-phonon coupling may be crucial to account for the insulating properties above T_C [6]. The presence of a Berry phase at large Hund-coupling also challenges predictions from the DE model [7]. In a series of recent papers (Refs. [8–12]) we have remarked that another phenomena occurring in manganites and not included in the DE description, namely the charge inhomogeneity, may be contained in a more fundamental Kondo-like model. More precisely, in those papers [8–10] it was

Springer Proceedings in Physics, Volume 85
Computer Simulation Studies in Condensed-Matter Physics XII
Eds.: D.P. Landau, S.P. Lewis and H.-B. Schüttler

reported the presence of *phase separation* (PS) between hole-undoped antiferromagnetic and hole-rich ferromagnetic regions in the low temperature phase diagram of the one-orbital FM Kondo model. In a recent paper by the same authors a similar phenomenon was also observed using two orbitals and Jahn-Teller phonons [11]. In this case the two phases involved are both spin ferromagnetic and the *orbital* degrees of freedom are responsible for the phase separation. Influence of the phase separation tendency on the nearby phases of the phase separation region in the temperature-density phase diagram was discussed and a detailed comparison between our calculations and experimental observations was made in a recent review paper by the present authors [12]. Upon the inclusion of long-range Coulombic repulsion, FM clusters embedded into AF background would be stabilized. Charge ordering in the form of non-trivial extended structures (such as stripes) could be possible, similarly as discussed for the cuprates [13–16] but now also including ferromagnetic domains. The analysis carried out by our group suggests that phenomena as rich as observed in the high-Tc superconductors may exist in the manganites as well, and hopefully our effort will induce further theoretical and experimental work in this context.

The present paper should be considered as an informal review of the present status of the computational studies of models for manganites which have reported the existence of phase separation at low temperatures. It does *not* pretend to be a comprehensive article, and thus we encourage the readers to consult the literature mentioned here to find out additional references, including our recent review paper [12].

The paper is organized as follows: in the next section the results for the case of the one-orbital Kondo model are reviewed, with emphasis on the phase diagram. Most of the calculations are performed with classical t_{2g}-spins. In the following section results recently reported for the case of the two-orbital model with Jahn-Teller phonons are described. Once again the main emphasis is given to the phase diagram. In both of those previous sections the main result is that tendencies to ferromagnetism and phase separation are in strong competition in these models. Such a phenomenon appears so clearly in all dimensions of interest and for such a wide range of models that it leads us to believe that it may be relevant for experiments in the context of manganites. A discussion of possible clustering behavior after Coulomb interactions are added are briefly described towards the end of the paper. In the last section conclusions are provided.

2 One-Orbital Ferromagnetic Kondo Model

The one-orbital FM Kondo model is defined as

$$H = -t \sum_{\langle \mathbf{ij} \rangle \sigma} (c^\dagger_{\mathbf{i}\sigma} c_{\mathbf{j}\sigma} + h.c.) - J_H \sum_{\mathbf{i}\alpha\beta} c^\dagger_{\mathbf{i}\alpha} \sigma_{\alpha\beta} c_{\mathbf{i}\beta} \cdot \mathbf{S}_{\mathbf{i}}, \tag{1}$$

where $c_{\mathbf{i}\sigma}$ are destruction operators for one species of e_g-fermions at site $\mathbf{i}$ with spin σ, and $\mathbf{S_i}$ is the total spin of the t_{2g} electrons, assumed localized. The first term is the e_g electron transfer between nearest-neighbor Mn-ions, $J_H > 0$ is the Hund coupling, the number of sites is L, and the rest of the notation is standard. The density is adjusted using a chemical potential μ. In this paper and Refs. [8,9] the spin $\mathbf{S_i}$ is considered classical (with $|\mathbf{S_i}| = 1$), unless otherwise stated. Although models beyond Eq.(1) may be needed to fully understand the manganites (notably those that include lattice effects as studied later in this paper), it is important to analyze in detail the properties of simple Hamiltonians to clarify if part of the experimental rich phase diagram can be accounted for using purely electronic models.

To study Eq. (1) in the t_{2g}-spin classical limit a Monte Carlo (MC) technique was used. The partition function of the model Eq. (1) with classical spins can be written as

$$Z = \prod_{\mathbf{i}}^{L} \left(\int_0^{\pi} d\theta_{\mathbf{i}} \sin\theta_{\mathbf{i}} \int_0^{2\pi} d\phi_{\mathbf{i}} \right) \mathrm{Tr_c}(\mathrm{e}^{-\beta(\mathrm{H}-\mu)}). \tag{2}$$

For a fixed configuration of angles $\{\theta_{\mathbf{i}}, \phi_{\mathbf{i}}\}$ the Hamiltonian amounts to non-interacting electrons moving in an external field. This problem can be diagonalized exactly since the Hamiltonian is quadratic in the fermionic variables. Once the $2L$ eigenvalues denoted by ϵ_λ are calculated, the resulting partition function becomes

$$Z = \prod_{\mathbf{i}}^{L} \left(\int_0^{\pi} d\theta_{\mathbf{i}} \sin\theta_{\mathbf{i}} \int_0^{2\pi} d\phi_{\mathbf{i}} \right) \prod_{\lambda=1}^{2L} (1 + e^{-\beta(\epsilon_\lambda - \mu)}). \tag{3}$$

The integrand in Eq.(3) is positive and, thus, a Monte Carlo simulation can be performed on the classical spin angles without having "sign" problems. This is the procedure followed in our study below. Physical quantities for the e_g electrons are obtained by first calculating "Green functions" for a fixed spin configuration $\{\theta_{\mathbf{i}}, \phi_{\mathbf{i}}\}$, *e.g.*,

$$\langle c_{i\sigma} c^\dagger_{j\sigma} \rangle = \frac{\mathrm{Tr_c}[\mathrm{c_{i\sigma}} \mathrm{c}^\dagger_{\mathrm{j}\sigma} \mathrm{e}^{-\beta(\mathrm{H}-\mu)}]}{\mathrm{Tr_c}(\mathrm{e}^{-\beta(\mathrm{H}-\mu)})} \tag{4}$$

$$= \sum_{\nu=1}^{2L} (\bar{T})_{i\sigma,\nu} \frac{1}{1 + e^{-\beta(\epsilon_\nu - \mu)}} (\bar{T}^\dagger)_{\nu, j\sigma}, \tag{5}$$

where $\bar{T}$ is the eigenmatrix of the Hamiltonian, and then carrying out a MC sampling for $\{\theta_{\mathbf{i}}, \phi_{\mathbf{i}}\}$.

The results are summarized in the phase diagram of Fig.1. In 1D (and also in 2D, not shown) and at low temperatures clear indications of (i) ferromagnetism, (ii) spin incommensurate (IC) correlations, (i) antiferromagnetism, and (vi) phase separation were identified. Results are also available in small

3D clusters and qualitatively they agree with those in Fig.1. The same occurs working at D $= \infty$ (see Refs. [8,9]). In 1D we also obtained results with quantum t_{2g}-spins S=3/2 which are in good agreement with Fig. 1[8,9].

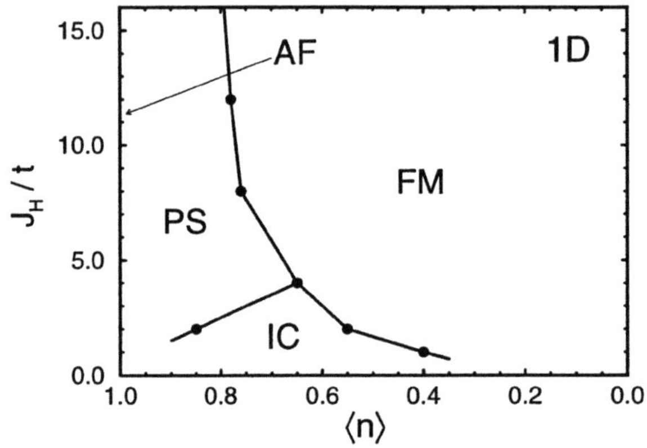

Fig. 1. Phase diagram of the FM Kondo model reported in Refs. [8,9] obtained using numerical techniques. FM, IC, AF, and PS denote regimes with FM correlations, spin incommensurate correlations, AF correlations, and with phase separation between undoped AF and hole-rich FM regions, respectively. The results shown correspond to one dimension and they were obtained with MC simulations at $T = t/75$ using chains with L=20, 30 and 40 sites

The boundaries of the FM region of the phase diagram were found evaluating the spin-spin correlation defined as $S(\mathbf{q}) = (1/L) \sum_{\mathbf{j},\mathbf{m}} e^{i(\mathbf{j}-\mathbf{m})\cdot\mathbf{q}} \langle \mathbf{S_j} \cdot \mathbf{S_m} \rangle$. Fig. 2(a) shows $S(\mathbf{q})$ at zero momentum vs. T/t for typical examples in 1D. The rapid increase of the spin correlations as T is reduced and as the lattice size grows clearly points towards the existence of ferromagnetism in the system [17]. It is natural to assume that the driving force for FM in this context is the DE mechanism. Repeating this procedure for a variety of couplings and densities, the robust region of FM shown in Fig.2(a) was determined. Fig.2(b) shows single particle excitation spectra $A(k,\omega)$ for the FM phase. The calculations were done on a chain of 40 sites at T=1/75 and the e_g electron density $\langle n \rangle = 0.5$[18]. In the figure, the dashed line is the dispersion relation for free spinless fermions, which is in excellent agreement with our numerical results.

In the small J_H/t region IC correlations were observed monitoring $S(\mathbf{q})$. Both in 1D and 2D there is one dominant peak which moves away from the AF location at $\langle n \rangle = 1$ towards zero momentum as $\langle n \rangle$ decreases. In the 2D clusters the peak moves from (π,π) towards $(\pi,0)$ and $(0,\pi)$, rather than along the main diagonal. Note that our computational study predicts IC correlations only in the small and intermediate J_H/t regime.

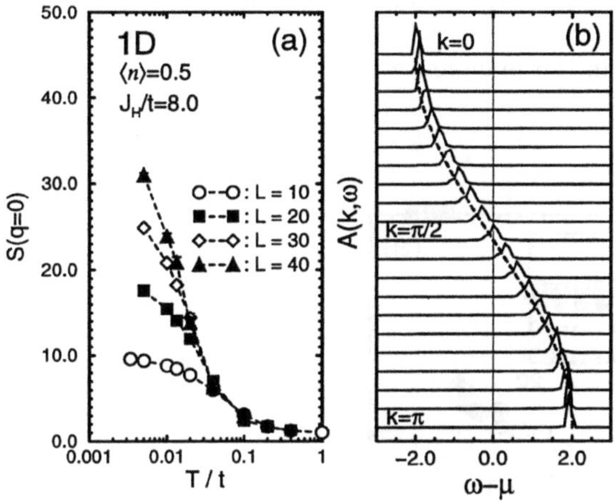

Fig. 2. (a) Spin-spin correlations of the classical spins at zero momentum $S(\mathbf{q}=0)$ vs temperature T. MC results for several lattice sizes are shown on chains. The density and coupling are shown. Closed shell boundary condition (BC) are used i.e. periodic BC (PBC) for $L = 10$ and 30 and antiperiodic BC (APBC) for $L = 20$ and 40. Results are taken from Ref. [8]. (b) Single particle excitation spectra $A(k,\omega)$ for the one orbital model with $J_H = 8t$ calculated on a chain of 40 lattices at $\langle n \rangle = 0.5$ and $T = t/75$ [10]. The dashed line is the dispersion relation for free spinless fermions, $-2t\cos k + \mu$

The main result of Refs. [8,9] is contained in Fig.3 where the computational evidence for the existence of phase separation in dimensions 1, 2, and ∞ is given. The presence of a discontinuity in $\langle n \rangle$ vs μ shows that some electronic densities can not be stabilized by tuning the chemical potential [19]. If the system is nominally prepared with such densities it will spontaneously separate into two regions with the densities corresponding to the extremes of the discontinuities of Fig.3. By analyzing these extremes the properties of the two domains can be studied. One region is undoped ($\langle n \rangle = 1$) with strong AF correlations, while the other contains all the holes and the spin-spin correlations between the classical spins are FM (see the inset of Fig.3a. The results are similar in D=2 and infinite). This is natural since holes optimize their kinetic energy in a FM background. On the other hand, at $\langle n \rangle = 1$ the DE mechanism is not operative: if the electrons fully align their spins they cannot move in the conduction band due to the Pauli principle. Then, energetically an AF pattern is formed. As J_H grows, the jump in Fig.3 is reduced and it tends to disappear in the $J_H = \infty$ limit.

Experimentally, PS may be detectable using neutron diffraction techniques, since $S(q)$ should present a two peak structure, one located at the AF position and the other at zero momentum. Since this also occurs in a canted ferromagnetic state care must be taken in the analysis of the experimental data. Another alternative is that Coulombic forces prevent the macroscopic accumulation of charge intrinsic of a PS regime. Thus, microscopic hole-rich domains may develop in the manganites, perhaps in the form of stripes.

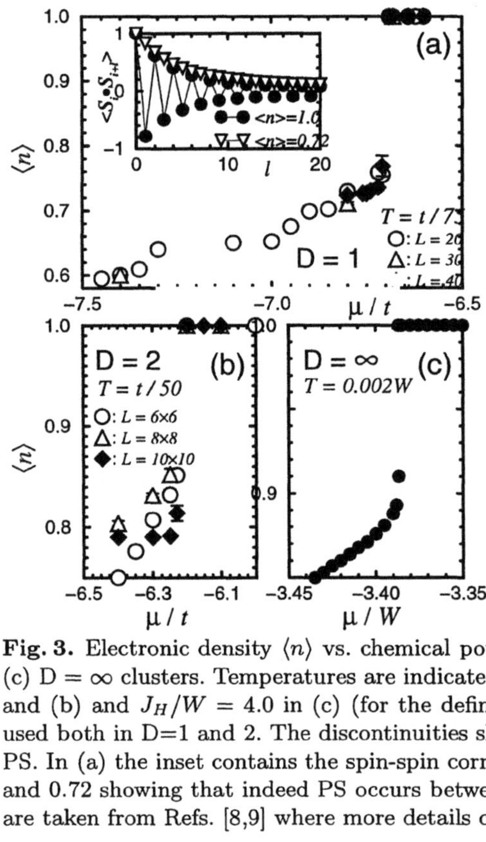

Fig. 3. Electronic density $\langle n\rangle$ vs. chemical potential μ in (a) D=1, (b) D=2, and (c) D $= \infty$ clusters. Temperatures are indicated. The coupling is $J_H/t = 8.0$ in (a) and (b) and $J_H/W = 4.0$ in (c) (for the definition of W see Ref. [8]). PBC were used both in D=1 and 2. The discontinuities shown in the figures are indicative of PS. In (a) the inset contains the spin-spin correlation in real space at densities 1.0 and 0.72 showing that indeed PS occurs between AF and FM regions. The results are taken from Refs. [8,9] where more details can be found

Although the phase diagram of Fig.1 has PS close to half-filling, actually this phenomenon also occurs at $\langle n\rangle \sim 0$ if an extra direct AF exchange interaction between the localized spins is included [10]. This coupling may be originated in a small hopping amplitude for the t_{2g} electrons. Fig.4 shows that at $\langle n\rangle = 0$, model Eq.(1) supplemented by a antiferromagnetic exchange coupling J'/t among the localized spins produces an AF phase, as in experiments, which upon electron doping induces a competition between AF (with no e_g-electrons) and FM electron-rich regions, similarly as for $\langle n \rangle = 1$ but replacing holes by electrons. Thus, PS or charge ordering could exist in manganites both at large and small fermionic densities. A careful study of the influence of the J' coupling on the phase diagram of the one orbital Kondo model has been presented in Ref. [10]

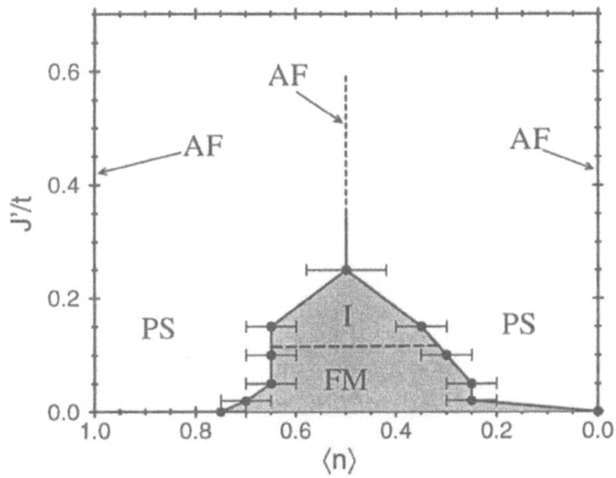

Fig. 4. The phase diagram for the one-dimensional FM Kondo model with antiferromagnetic exchange coupling J' between the localized t_{2g} spins. Here Hund's coupling $J_H = 8t$ and the t_{2g} spins are treated classical. AF indicates the antiferromagnetic phase, I the insulating region, FM the ferromagnetic metallic phase, and PS denotes phase separation. The results are taken from Ref. [10]

It should be mentioned here why phase separation occurs at $\langle n \rangle \sim 0$, as well as $\langle n \rangle \sim 1$. In Fig. 5(a) the chemical potential μ vs. electron density $\langle n \rangle$ for the one dimensional FM Kondo model with AF exchange coupling between the classical spins is presented. It clearly shows that there are forbidden densities close to 0 and to 1 at the critical chemical potentials $\mu_c \sim -9.4$ and $\mu_c' \sim -7.0$, respectively, indicating that states with these densities are phase separated. Let us now try to introduce electrons into the AF state with $\langle n \rangle = 0$ by increasing μ. Up to the critical chemical potential μ_c, the state remains AF. The density of states $N(\omega)$ for the AF state at $\mu = \mu_c - 0^+$ is calculated in Fig. 5(b) (solid line). At $\mu = \mu_c$, electrons can be introduced into the system. Once electrons are added into the system, the electronic state is changed drastically from AF to FM states because of the large gain of kinetic energy. $N(\omega)$ for the FM state at $\mu = \mu_c + 0^+$ is presented in Fig.

5(b) (dashed line). The important fact is that the bandwidth of the FM state is much larger that the one of the AF state, and it causes the jump of $\langle n \rangle$ at the critical chemical potential μ_c. The same occurs at $\langle n \rangle \sim 1$, as shown in Fig. 5(c).

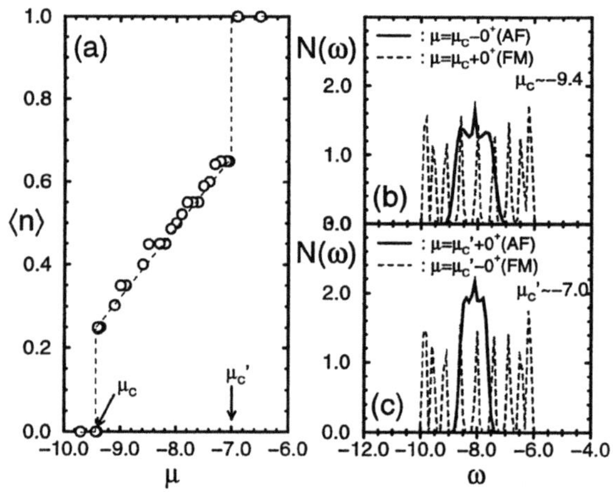

Fig. 5. (a) μ vs. $\langle n \rangle$ obtained with the MC technique applied to the one dimensional FM Kondo model with classical spins. Here $J_H = 8t$, $J' = 0.05t$, $T = 0.01t$, and $L = 20$ are used. It shows clear jumps of $\langle n \rangle$ at $\mu \sim -9.4(\mu_c)$ and $\mu \sim -7.0(\mu_c')$. The dashed line is a guide to the eyes. (b) Density of states $N(\omega)$ for the AF (solid line) and FM (dashed line) states at $\mu = \mu_c - 0^+$ and $\mu = \mu_c + 0^+$, respectively. The parameters are the same as in (a). (c) Same as (b) but using different chemical potential, $\mu = \mu_c' + 0^+$(solid line, AF state) and $\mu_c' - 0^+$(dashed line, FM state). The spike structure of DOSs for FM states in (b) and (c) is due to finite size effects

3 Two-Orbital FM Kondo Model with Jahn-Teller Phones

In spite of the rich phase diagram observed in the study of the one orbital FM Kondo Model discussed in the previous section and its similarities with experiments, there are aspects of the real manganites that require a more sophisticated approach. For instance, dynamical Jahn-Teller (JT) distortions are claimed to be very important [6], and a proper description of the recently observed orbital order [20] obviously needs at least two orbitals. Such a multi-orbital model with JT phonons is nontrivial, and thus far it has been studied only using the dynamical mean-field approximation [21]. The previous experience with the one orbital case suggests that a computational analysis is actually crucial to understand its properties. In addition, it is conceptually interesting to analyze whether the PS described before [8–10] exists also in multi-orbital models.

The first computational study of a two orbital model for manganites including the Jahn-Teller electron-phonon coupling was reported recently by the authors in Ref. [11] and here a summary of the main conclusions will be discussed. The results show a rich phase diagram including a novel regime of PS induced by the *orbital*, rather than the spin, degrees of freedom. The Hamiltonian used in that study had three contributions

$$H_{KJT} = H_K + H_{JT} + H_{AF}. \tag{6}$$

The first term is

$$H_K = -\sum_{\langle \mathbf{ij}\rangle \sigma ab} t_{ab}(c^\dagger_{\mathbf{i}a\sigma}c_{\mathbf{j}b\sigma} + h.c.) - J_H \sum_{\mathbf{i}a\alpha\beta} \mathbf{S}_\mathbf{i} \cdot c^\dagger_{\mathbf{i}a\alpha}\sigma_{\alpha\beta}c_{\mathbf{i}a\beta}, \tag{7}$$

where $\langle\mathbf{ij}\rangle$ denotes nearest-neighbor lattice sites, $J_H > 0$ is the Hund coupling, the hopping amplitudes t_{ab} are described in Ref. [22], the t_{2g} spins $\mathbf{S}_\mathbf{i}$ are assumed to be classical (with $|\mathbf{S}_\mathbf{i}| = 1$) since their actual value in Mn-oxides (3/2) is large [23], and the rest of the notation is standard. None of the results described below depends crucially on the set $\{t_{ab}\}$ selected [22]. The energy units are chosen such that $t_{11} = 1$ in the x-direction. In addition, since J_H is large in the real manganites, here it will be fixed to 8 unless otherwise stated. The e_g-density $\langle n\rangle$ is adjusted using a chemical potential μ.

The coupling with JT-phonons is through [6,24,25]

$$H_{JT} = \lambda \sum_{\mathbf{i}ab\sigma} c^\dagger_{\mathbf{i}a\sigma}Q^{ab}_\mathbf{i}c_{\mathbf{i}b\sigma} + \frac{1}{2}\sum_\mathbf{i}({Q^{(2)}}^2_\mathbf{i} + {Q^{(3)}}^2_\mathbf{i}), \tag{8}$$

where $Q^{11}_\mathbf{i} = -Q^{22}_\mathbf{i} = Q^{(3)}_\mathbf{i}$, and $Q^{12}_\mathbf{i} = Q^{21}_\mathbf{i} = Q^{(2)}_\mathbf{i}$. These phonons are assumed to be classical, which substantially simplifies the computational study. This is a reasonable first approximation towards the determination of the phase diagram of the H_{KJT} model. Finally, a small antiferromagnetic coupling between the t_{2g}-spins is needed to account for the AF character of the real materials even when all La is replaced by Ca or Sr (fully doped manganites). This classical Heisenberg term is $H_{AF} = J'\sum_{\langle\mathbf{ij}\rangle}\mathbf{S}_\mathbf{i}\cdot\mathbf{S}_\mathbf{j}$, where J' is fixed to 0.05 throughout the paper, a value compatible with experiments [26]. To study H_{KJT} a Monte Carlo (MC) technique similar to that employed in Refs. [8–10] and in the previous section for the one orbital problem was used. Finally, to analyze orbital correlations the pseudospin operator $\mathbf{T}_\mathbf{i} = \frac{1}{2}\sum_{\sigma ab} c^\dagger_{\mathbf{i}a\sigma}\sigma_{ab}c_{\mathbf{i}b\sigma}$ was used, while for spin correlations the operator is standard. The Fourier transform of the pseudospin correlations is defined as $T(\mathbf{q}) = \frac{1}{L}\sum_{\mathbf{l},\mathbf{m}} e^{i\mathbf{q}\cdot(\mathbf{l}-\mathbf{m})}\langle\mathbf{T}_\mathbf{m}\cdot\mathbf{T}_\mathbf{l}\rangle$, with a similar definition for the spin structure factor $S(\mathbf{q})$.

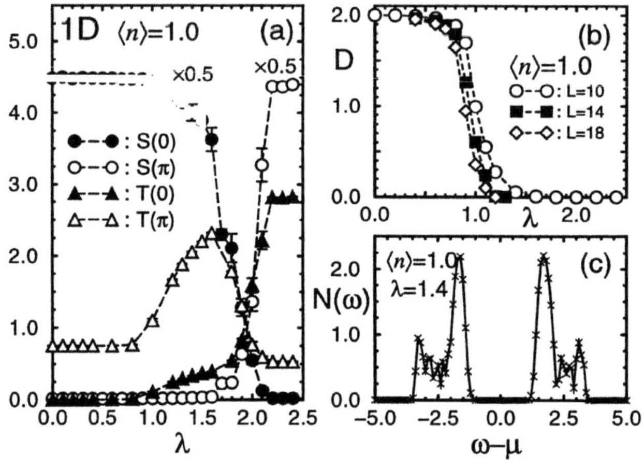

Fig. 6. (a) $T(q)$ and $S(q)$ vs λ, working at $\langle n\rangle = 1.0$, $T = 1/75$, $J_H = 8$, $J' = 0.05$, and in 1D with 10 sites. $\{t_{ab}\}$ correspond to set T_1 (see [22]). Results with sets T_2 and T_3 are qualitatively the same; (b) Drude weights D as a function of λ on various chains indicated in the figure. The rest of the parameters are the same as in (a); (c) Density of states $N(\omega)$ for $\lambda = 1.4$, $\langle n\rangle = 1.0$, $T = 1/75$, $J_H = 8$, $J' = 0.05$, and in 1D with 20 sites with $\{t_{ab}\}$ correspond to set T_1

Let us first consider the limit $\langle n\rangle = 1.0$, corresponding to undoped manganites. Fig.6(a) shows $T(q)$ and $S(q)$ at representative momenta $q = 0$ and π vs. λ for 1D. For small electron-phonon coupling the results are similar to those at $\lambda = 0.0$, namely a large $S(0)$ indicates a tendency to spin-FM order induced by DE mechanism (as in the qualitatively similar one orbital problem at $\langle n\rangle = 0.5$ [8]). The small values of $T(q)$ imply that in this regime the orbitals remain disordered. When the coupling reaches $\lambda_{c1} \sim 1.0$, the rapid increase of $T(\pi)$ now suggests that the ground state has a tendency to form a *staggered* (or "antiferro" AF) orbital pattern, with the spins remaining FM aligned since $S(0)$ is large. The existence of this phase was discussed before, but using multi-orbital Hubbard models with Coulomb interactions and without phonons [27]. Our results show that it can be induced by JT phonons as well. As the coupling increases further, another transition at $\lambda_{c2} \sim 2.0$ occurs to a spin-AF orbital-FM state ($S(\pi)$ and $T(0)$ are large). In this region a one orbital approximation should be suitable.

Let us discuss the electronic properties in the three regimes at $\langle n\rangle = 1$. The algorithm used in Ref. [8–10] allows us to calculate *real-time* dynamical responses accurately, including the optical conductivity $\sigma(\omega > 0)$, since all the eigenvectors in the fermionic sector for a given spin and phonon configuration are obtained exactly. From the sum-rule, the Drude-weight D was calculated and as shown in Fig. 6(b) D vanishes at λ_{c1} signaling a *metal-insulator* transition. Here the insulating phase is spin-FM and orbital-AF, while the metallic one is spin-FM and orbital-disordered. The density of states (DOS) for $\lambda > \lambda_{c1}$ was also calculated in Ref. [11] and Fig. 6(c) and it shows a clear gap at the Fermi level. The qualitative shape of D vs λ on 4^2 and 4^3 clusters was found to be the same as in 1D and, thus, it is reasonable to assume that the metal-insulator transition exists also in all dimensions of interest.

Consider now the influence of hole doping on the $\langle n \rangle = 1.0$ phase diagram, with special emphasis on the stability of other densities as μ is varied. Fig.7 shows $\langle n \rangle$ vs μ in the intermediate-λ regime. It is remarkable that *two* regions of unstable densities exist at low-T (similar conclusions were reached using the Maxwell's construction [28]). These instabilities signal the existence of PS in the H_{KJT} model. At low-density there is separation between an (i) empty e_g-electron band with AF-ordered t_{2g}-spins and a (ii) metallic spin-FM orbital uniformly ordered phase. This regime of PS is analogous to the spin-driven PS found at low-density in the one orbital problem [10]. On the other hand, the unstable region near $\langle n \rangle = 1.0$ is *not* contained in the one orbital case. Here PS is between the phase (ii) mentioned above, and (iii) the insulating intermediate-λ spin-FM and orbital-AF phase described in Fig.6(a) [29]. In 2D the results were found to be very similar (Ref. [11]). The driving force for this novel regime of PS are the *orbital* degree of freedom, since the spins are uniformly ordered in both phases involved. Studying $\langle n \rangle$ vs μ, for $\lambda < \lambda_{c1}$ only PS at small densities is observed, while for $\lambda > \lambda_{c2}$ the PS close to $\langle n \rangle = 1$ is similar to the same phenomenon observed in the one orbital problem since it involves a spin-AF orbital-FM phase [8].

The phase diagram of the 1D H_{KJT} model is given in Fig. 8. The two PS regimes are shown, together with the metallic spin-FM region. This phase is separated into two regions, one ferro-orbital ordered and the other orbitally disordered. The existence of these two regimes can be deduced from the behavior of the pseudospin correlations, the mean value of the pseudospin operators, and the probability of double occupancy of the same site with different orbitals. The results are similar for several $\{t_{ab}\}$ sets [22]. The simulations of Ref. [11] suggest that the qualitative shape of Fig.8 should be valid also in $D = 2$ and 3.

The macroscopic separation of two phases with different densities, and thus different charges, should actually be prevented by long-range Coulombic interactions, which were not incorporated into the one- and two-orbital models discussed thus far. Even including screening and polarization effects, a complete separation leads to a huge energy penalty. This finding immediately suggests that the two large regions involved in the process will break into smaller pieces to spread the charge more uniformly, leading to a stable state with clusters of one phase embedded in the other (see also Ref. [30]). It is expected that the competition between the attractive DE tendencies among carriers and the Coulomb forces will determine the size and shape of the resulting clusters. Either sizable droplets or polarons may arise

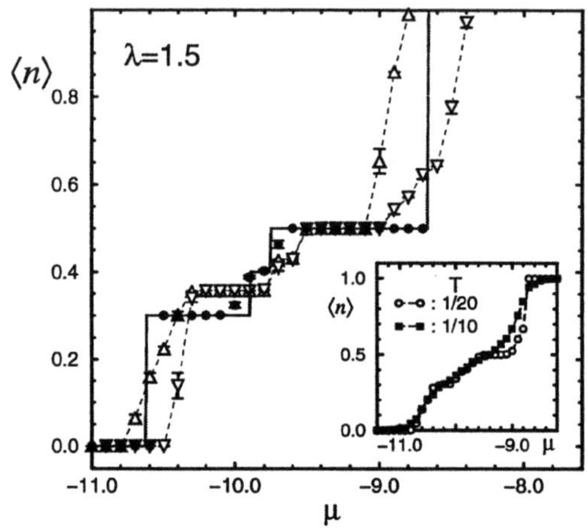

Fig. 7. $\langle n \rangle$ vs μ at $\lambda = 1.5$, $L = 10$, and $T = 1/40$ (solid circles). The discontinuities near $\langle n \rangle = 1.0$ and 0.0 show the existence of unstable densities. The solid line is obtained from the Maxwell's construction. The triangles are results also at $\lambda = 1.5$ and $T = 1/40$, but using 14 sites and only 2×10^4 MC sweeps to show the appearance of *hysteresis* loops as in a first-order transition. The inset shows the T-dependence of the results at $L = 10$. These results were taken from Ref. [11]

as the most likely configuration [30]. However, the ideas presented here will still be described as the "PS scenario," with the understanding that only microscopic phase segregation is the resulting net effect of the DE-Coulomb competition. Related ideas have been previously discussed in the context of the cuprates [31], with attractive interactions generated by antiferromagnetism or phonons. An exception to the existence of only purely microscopic effects occurs if the competing phases have approximately the same density, as observed experimentally at x = 0.5. In this case, large-scale PS can be expected.

Phase separation in manganese oxides has clear similarities with the previously discussed charge inhomogeneities observed in copper and nickel oxides [31]. Actually, studies of 1D generalizations of the t-J model by Riera et al. [32] showed that as the localized spin magnitude S grows, the phase diagram is increasingly dominated by either FM or PS tendencies. The importance of PS arises from the dominance of the Heisenberg interactions over the kinetic energies as S increases, which causes holes to be expelled from the AF-regions because they damage the spin environment. The tendency toward phase segregation decreases across the transition-metal-row, from a strong tendency in Mn, to a weak tendency in Cu [32].

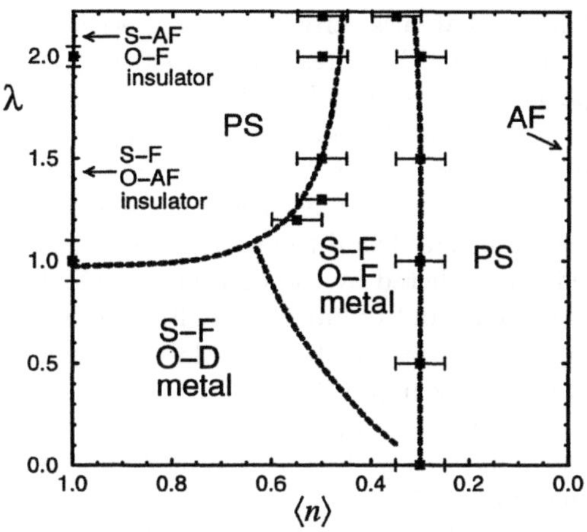

Fig. 8. Phase diagram of the H_{KJT} model at low-T, $J_H = 8.0$, $J' = 0.05$, and using set T_1 for $\{t_{ab}\}$. S − F and S − AF denote regimes with FM- and AF-spin orders, respectively. O − D, O − F, and O − AF represent states with disordered, uniform and staggered orbital orders, respectively. PS means phase separation. These results were taken from Ref. [11]

4 Conclusions

A variety of recent calculations have found PS tendencies in models for manganites, usually involving FM and AF phases. These tendencies, which should lead to a stable but microscopically inhomogeneous state upon the inclusion of Coulomb interactions, compete strongly with ferromagnetism in the phase diagrams and are expected to increase substantially the resistivity. Particularly, when two-orbital models are studied, the results are in good agreement with a large list of experimental observations as reviewed by the present authors recently [12]. Tendencies toward charge inhomogeneous states exist in real manganese oxides all around the FM-phase in the temperature-density phase diagram. The computer simulations have shown that the region with PS tendencies substantially influences the stable FM-phase by increasing its compressibility, an aspect that can be tested experimentally. This provides a rationalization for the experimental observation of a large MR effect at the boundaries of the FM-phase. The presence of short-range charge correlations is certainly a crucial feature of the PS scenario.

However, considerable work still remains to be done. The inclusion of extended Coulomb interactions and the stabilization of the $x = 0.5$ charge-ordered state are the next challenges for computational studies. Analytical techniques beyond the local mean-field approximations are needed to capture the essence of the charge inhomogeneous state. Macroscopic phenomenological approaches should be used to obtain predictions for transport properties and the shapes of the clusters that arise from the competition of the DE attraction and Coulomb repulsion. These resulting clusters are surely not static but fluctuating, specially above T_C. In related problems of nuclear physics at high density, several geometries have been found including spherical drops, rod-like structures (stripes [31] or "spaghetti") and plate-like ones ("lasagna") [33]. Similar rich phenomena may occur in manganites. On the experimental front it is crucial to establish if the various regimes with charge inhomogeneities are related, as predicted by the theoretical calculations [12]. For example, work should be carried out to link the small x regime of (La,Ca)-manganites where FM-droplets appear, with the polarons reported at the x=1/3 density, and beyond into the highly hole-doped regime. In addition, phase segregation tendencies should also be studied close to the fully doped limit $\langle n \rangle \ll 1$ of manganese oxides, as well as in related compounds such as dilute magnetic semiconductors [34,35].

E. D. and A. M. are supported by the NSF grant DMR-9520776. S. Y. thanks the NHMFL for support.

References

1. Y. Tokura et al., J. Appl. Phys. **79**, 5288 (1996); A. P. Ramirez, J. Phys.: Condens. Matter **9**, 8171 (1997); Y. Tokura, in *Colossal Magnetoresistance Oxides*, ed. Y. Tokura, Gordon & Breach, Monographs in Cond. Matt. Science; J. M. D. Coey, M. Viret, and S. von Molnar, *Mixed-valence Manganites*, Adv. Phys. 1999, in press.
2. P. E. Schiffer, A. P. Ramirez, W. Bao, and S-W. Cheong, Phys. Rev. Lett. **75**, 3336 (1995); A. P. Ramirez et al., Phys. Rev. Lett. **76**, 3188 (1996); C. H. Chen and S-W. Cheong, Phys. Rev. Lett. **76**, 4042 (1996); S-W. Cheong and C. H. Chen, in *Colossal Magnetoresistance and Related Properties*, ed. by B. Raveau and C. N. R. Rao (World Scientific, 1998).
3. C. Zener, Phys. Rev. **82**, 403 (1951).
4. P. W. Anderson and H. Hasegawa, Phys. Rev. **100**, 67 (1955).
5. P. G. de Gennes, Phys. Rev. **118**, 141 (1960).
6. A. J. Millis, et al., Phys. Rev. Lett. **74**, 5144 (1995); H. Röder, et al., Phys. Rev. Lett. **76**, 1356 (1996).
7. E. Müller-Hartmann and E. Dagotto, Phys. Rev. **B 54**, R6819 (1996).
8. S. Yunoki, J. Hu, A. Malvezzi, A. Moreo, N. Furukawa, and E. Dagotto, Phys. Rev. Lett. **80**, 845 (1998).
9. E. Dagotto, S. Yunoki, A. Malvezzi, A. Moreo, J. Hu, S. Capponi, D. Poilblanc, and N. Furukawa, Phys. Rev. **B 58**, 6414 (1998).
10. S. Yunoki and A. Moreo. Phys. Rev. **B58**, 6403 (1998).
11. S. Yunoki, A. Moreo, and E. Dagotto, Phys. Rev. Lett. **81**, 5612 (1998).
12. A. Moreo, S. Yunoki, and E. Dagotto, cond-mat/9901057 (to appear in Science).
13. V. J. Emery, S. A. Kivelson, and H. Q. Lin, Phys. Rev. Lett. **64**, 475 (1990). See also V. J. Emery, and S. A. Kivelson, Physica **C 209**, 597 (1993).
14. E. Dagotto, Rev. Mod. Phys. **66**, 763 (1994), and references therein.
15. J. M. Tranquada et al., Nature **375**, 561 (1995), and references therein.

16. U. Löw et al., Phys. Rev. Lett. **72**, 1918 (1994); S. Haas et al., Phys. Rev. **B 51**, 5989 (1995).
17. In both 1D and 2D the Mermin-Wagner theorem forbids a nonzero T_C in the model. However, the correlation lengths can be very large even at finite T/t.
18. All one needs to do is just do a Monte Carlo sampling for the classical spins of $\sum_{\mu} | \sum_{i,\sigma} e^{ik \cdot r_i} (\bar{T})_{i\sigma,\mu} |^2 \delta(\omega - \epsilon_{\mu})$.
19. S-Q. Shen and Z. D. Wang, Phys. Rev. **B 58**, R8877 (1998); D. Arovas and F. Guinea, Phys. Rev. **B 58**, 9150 (1998); D. Arovas, G. Gomez-Santos, and F. Guinea, cond-mat/9805399; H. Yi and J. Yu, Phys. Rev. **B 58**, 11123 (1998); M. Yamanaka, W. Koshibae, and S. Maekawa, Phys. Rev. Lett. **81**, 5604 (1998); M. Yu. Kagan, et al., cond-mat/9804213; L-J. Zou, Q-Q. Zheng, and H. Q. Lin, cond-mat/9806015.
20. Y. Murakami et al., Phys. Rev. Lett. **80**, 1932 (1998); **81**, 582 (1998).
21. A. J. Millis et al., Phys. Rev. **B 54**, 5405 (1996).
22. Most of the work in one-dimension (1D) has been performed using $t_{11} = t_{22} = 2t_{12} = 2t_{21}$ (set T_1), but results have also been obtained with $t_{11} = t_{22}$ and $t_{12} = t_{21} = 0$ (T_2), as well as with the hopping that takes into account the proper orbital overlap, namely $t_{11} = 3t_{22} = \sqrt{3}t_{12} = \sqrt{3}t_{21}$ (T_3) (see S. Ishihara et al., Phys. Rev. **B 56**, 686 (1997)). In two-dimensions (2D), the set T_1 in both directions was used, but also the combination of T_3 in the y-direction and $t_{11} = 3t_{22} = -\sqrt{3}t_{12} = -\sqrt{3}t_{21}$ (T_4) in the x-direction. Finally, in three-dimensions (3D) T_4 was used in the x-direction, T_3 in the y-direction, and $t_{11} = t_{12} = t_{21} = 0$, $t_{22} = 4/3$ (T_5) in the z-direction.
23. This approximation was shown to be accurate in Ref. [8,9].
24. J. Kanamori, J. Appl. Phys. (Suppl.) **31**, 145 (1960).
25. J. B. Goodenough, *Magnetism and the Chemical Bond* (John Wiley & Sons, New York, 1963).
26. T. G. Perring et al., Phys. Rev. Lett. **78**, 3197 (1997).
27. J. B. Goodenough, Phys. Rev. **100**, 565 (1955); K. I. Kugel and D. I. Khomskii, JETP Lett. **15**, 446 (1972); S. Ishihara, J. Inoue and S. Maekawa, Phys. Rev. **B 55**, 8280 (1997); T. Mizokawa and A. Fujimori, Phys. Rev. **B 56**, R493 (1997).
28. The energies needed for this construction are obtained at low-T from the MC evolution of the density [10].
29. Slave-fermion mean-field studies of a model with strong Coulomb interactions and no JT phonons by S. Maekawa et al. (see, e.g. cond-mat/9902266) have also recently observed a similar PS pattern.
30. Some numerical results already support this picture: A. Malvezzi, S. Yunoki, and E. Dagotto, cond-mat/9809281, to appear in PRB.
31. K. A. Müller, and G. Benedek, eds., proceed. of the conf. *Phase Separation in Cuprate Superconductors*, World Scientific (1993); V. J. Emery and S. A. Kivelson, Physica C **209**, 597 (1993); C. Castellani, C. Di Castro, and M. Grilli, Phys. Rev. Lett. **75**, 4650 (1995); J. M. Tranquada et al., Nature **375**, 561 (1995); E. Dagotto et al., Phys. Rev. **B 49**, 3548 (1994); and references therein. See also L. P. Gor'kov and A. Sokol, JETP Lett. **46**, 420 (1987).
32. J. Riera, K. Hallberg, and E. Dagotto, Phys. Rev. Lett. **79**, 713 (1997).
33. See C. P. Lorenz, D. G. Ravenhall, and C. J. Pethick, Phys. Rev. Lett. **70**, 379 (1993), and references therein. The authors thank E. Fradkin for bringing these references to our attention.

34. T. Kasuya and A. Yanase, Rev. Mod. Phys. **40**, 684 (1968); S. von Molnar and S. Methfessel, J. Appl. Phys. **38**, 959 (1967); H. Ohno, et al., Phys. Rev. Lett. **68**, 2664 (1992); H. Ohno, Science **281**, 951 (1998).
35. J. L. Gavilano, B. Ambrosini, P. Vonlanthen, H. R. Ott, D. P. Young, and Z. Fisk, Phys. Rev. Lett. **81**, 5648 (1998).

Part II

Phase Transitions and Critical Phenomena

Critical Finite-Size Scaling with Constraints: Fisher Renormalization Revisited

M. Krech

Institut für Theoretische Physik, RWTH Aachen, 52056 Aachen, Germany

Abstract. The influence of a thermodynamic constraint on the critical finite-size scaling behavior of three-dimensional Ising and XY models is analyzed by Monte-Carlo simulations. Within the Ising universality class constraints lead to Fisher renormalized critical exponents, which modify the asymptotic form of the scaling arguments of the universal finite-size scaling functions. Within the XY universality class constraints lead to very slowly decaying corrections inside the scaling arguments, which are governed by the specific heat exponent α. If the modification of the scaling arguments is properly taken into account in the scaling analysis of the data, finite-size scaling functions are obtained, which are *independent* of the constraint as anticipated by analytic theory.

1 Introduction

The theoretical investigation of classical spin systems has played a key role in the understanding of phase transitions, critical behavior, scaling, and universality [1,2]. In particular, the classical Ising, the XY, and the Heisenberg model are the most relevant spin models in three dimensions. Each of these simple models represents a universality class which, apart from the spatial dimensionality and the range of the interactions, is characterized by the number of components of the order parameter , e.g, the magnetization in the case of ferromagnetic models. Real systems, however, suffer from various kinds of imperfections, e.g., lattice defects, impurities, or vacancies. In an experiment, which is designed to probe critical behavior as a function of temperature, the presence of, say, impurities on the lattice constitutes a thermodynamic constraint, because in a given sample the impurity concentration will remain constant during the temperature scans. According to the concepts of thermodynamics the impurity concentration n_i can be written as the derivative of the grand canonical potential with respect to the chemical potential μ_i of the impurities, where other parameters like the temperature and the volume of the system are kept fixed. Now the question arises how the critical singularities in the grand canonical potential are affected when the thermodynamic ensemble is changed from 'fixed μ_i' to 'fixed n_i', where the location of the critical temperature T_c depends on the particular values of μ_i or n_i, respectively. The answer to this question has been given a long time ago by Michael Fisher [3]. Provided, that the critical singularites have their usual form in the 'fixed

Springer Proceedings in Physics, Volume 85
Computer Simulation Studies in Condensed-Matter Physics XII
Eds.: D.P. Landau, S.P. Lewis and H.-B. Schüttler

μ_i' ensemble, then the constraint $n_i = const.$ amounts to a reparameterization of the reduced temperature $t = (T - T_c(n_i))/T_c(n_i)$ of the *constrained* system in terms of the reduced temperature $\tau = (T - T_c(\mu_i))/T_c(\mu_i)$ of the *unconstrained* system according to [3]

$$t = a\tau + b\tau|\tau|^{-\alpha} + \dots, \tag{1}$$

where a and b are nonuniversal constants and the dots indicate higher order contributions. Apart from a linear term (1) contains a singular contribution which is characterized by the critical exponent $1 - \alpha$ of the entropy density . Which of the two terms in (1) is the leading one for $t, \tau \to 0$ depends on the sign of α. Within the Ising universality class in $d = 3$ dimensions $\alpha \simeq 0.109$ [4] so that $|\tau| \sim |t|^{1/(1-\alpha)}$ to leading order and therefore the critical exponents β (order parameter), γ (susceptibility), and ν (correlation length) of the unconstrained system undergo 'Fisher renormalization' in the constrained system according to [3]

$$\beta \to \beta' = \beta/(1-\alpha), \quad \gamma \to \gamma' = \gamma/(1-\alpha), \quad \nu \to \nu' = \nu/(1-\alpha). \tag{2}$$

The specific heat exponent α requires a more careful analysis, because the specific heat is the temperature derivative of the entropy which in addition to the 'renormalization' displayed in (2) causes a sign change

$$\alpha \to \alpha' = -\alpha/(1-\alpha). \tag{3}$$

Note that analytic background contributions to the entropy of the unconstrained system become *singular* in the constrained system due to the singularity in the reparameterization given by (1).

Within the XY universality class in $d = 3$ the exponent α is negative [4], where probably the best current estimate $\alpha \simeq -0.013$ is obtained from an experiment on ^{4}He near the superfluid transition [5]. For negative α the linear term on the r.h.s. of (1) is the dominating one for $\tau \to 0$. However, the XY universality class α is so small, that in practice the singular term in (1) can never be neglected. Instead, the singular contribution to (1) gives rise to very slowly decaying correction terms which must not be confused with Wegner corrections to scaling . These correction terms have to be considered in any scaling analysis in order to obtain correct values for the critical exponents.

If the system is finite, which is neccessarily the case for any Monte - Carlo simulation, all critical singularities are rounded, i.e., all quantities are analytic functions of the thermodynamic parameters [6] so that a thermal singularity as shown in (1) does not occur. Critical finite-size rounding effects in, e.g., a cubic box L^d are captured by *universal* finite-size scaling functions [6,7] which restore all critical singularities in the limit $L \to \infty$. Following the line of argument in [3], (1) then has to be replaced by

$$t = a\tau + \tau|\tau|^{-\alpha} f(\tau L^{1/\nu}), \tag{4}$$

where $f(x)$ is the finite-size scaling function of the entropy density and $x = \tau L^{1/\nu}$ is a convenient choice of its scaling argument. For $\tau \to 0$ at finite

L the singular prefactor of $f(x)$ in (4) must be cancelled so that one has $f(x) = A|x|^\alpha + \ldots$ in the limit $x \to 0$, where A is a nonuniversal constant such that $f(x)/A$ is a *universal* function of its argument. To leading order in τ the reparameterization of the reduced temperature t of the constrained system is therefore *linear* in the reduced temperature τ of the unconstrained system and one finds

$$t = \tau(a + AL^{\alpha/\nu}). \tag{5}$$

According to (5) the finite-size scaling argument x in the constrained system is given by

$$x = \tau L^{1/\nu} = tL^{1/\nu}/(a + AL^{\alpha/\nu}), \tag{6}$$

where the *shape* of the finite-size scaling functions is maintained [8], i.e., the presence of the constraint *only* affects the form of the scaling argument x. For $\alpha > 0$ (6) asymptotically reduces to $x = tL^{1/\nu'}/A$ for large L in accordance with Fisher renormalization (see (2)). For $\alpha < 0$ (6) captures the aforementioned slowly decaying corrections to the asymptotic critical behavior in the XY universality class when a thermodynamic constraint is present. Note that $A > 0$ for the Ising universality class and that $A < 0$ for the XY universality class.

In the remainder of this paper a simple spin model is introduced which can be efficiently simulated with existing Monte - Carlo algorithms both with and without constraints in three dimensions. For the Ising and the XY version of the model finite-size scaling according to (6) is tested for the modulus of the order parameter, the susceptibility, and the specific heat.

2 Model and Simulation Method

The model system which is investigated here can be described as an $O(N-1)$ symmetric classical 'planar' ferromagnet in a transverse magnetic field. The model Hamiltonian reads

$$\mathcal{H} = -J \sum_{\langle ij \rangle} \sum_{x=1}^{N-1} S_i^x S_j^x - h \sum_i S_i^N, \tag{7}$$

where $\langle ij \rangle$ denotes a nearest neighbor pair of spins on a simple cubic lattice in $d = 3$ dimensions. The lattice contains L lattice sites in each direction and in order to avoid surface effects periodic boundary conditions are applied. Each spin $\mathbf{S}_i$ is a classical spin with N components $\boldsymbol{S}_i = (S_i^1, S_i^2, \ldots, S_i^N)$ with the normalization $|\boldsymbol{S}_i| = 1$ for each lattice site i. The magnetic field h in (7) only acts on the N-components of the spins which are not coupled by the exchange interaction J. From the symmetry of the Hamiltonian it is obvious, that the model belongs to the $O(N-1)$ universality class in $d = 3$, where nonuniversal quantities like the critical temperature $T_c = T_c(h)$ depend on the strength of the transverse field h. Note that $T_c(h)$ is symmetric around $h = 0$ and decreases with increasing h, because the spins become more and more aligned with the N-direction as $\pm h$ is increased and due to the normalization condition the typical interaction energy between pairs of spins is decreased.

The Hamiltonian given by (7) defines the unconstrained model. The constraint is imposed on the transverse magnetization M in the form

$$M \equiv \sum_i S_i^N = const., \tag{8}$$

where the Hamiltonian of the constrained model is given by

$$\mathcal{H}_M = -J \sum_{\langle ij \rangle} \sum_{x=1}^{N-1} S_i^x S_j^x = \mathcal{H} + hM. \tag{9}$$

The critical temperature of the constrained model is a symmetric and monotonically decreasing function of the prescribed transverse magnetization M. The transverse field h here plays the part of the chemical potential μ_i of impurities (see Sect. 1) and the transverse magnetization M accordingly plays the part of the impurity concentration n_i. It is also possible to implement $O(N)$ spin models with impurities or vacancies with diffusion in order to mimic the situation discussed in [3]. However, the fact that the Hamiltonians given by (7) and (9) only require a single 'species' with a single coupling constant leads to some simplifications in the algorithms. Note that the symmetric constrained model ($M = 0$) becomes equivalent to the symmetric unconstrained model ($h = 0$) for sufficiently large lattices. In particular, both versions of the symmetric model have the same T_c.

The Monte-Carlo algorithm is chosen as a hybrid scheme, where each hybrid Monte-Carlo step consists of 10 updates each of which can be one of the following: one Metropolis sweep of the whole lattice, one single cluster Wolff update [9], or one overrelaxation update of the whole lattice [10], where the latter can only be applied for $N \geq 3$. The Metropolis algorithm updates the lattice sequentially and works in the standard way for the unconstrained model. For the constrained model the constraint $M = const.$ is observed locally by applying a Kawasaki update dynamics for the Nth components of the spins. For each lattice site i a nearest neighbor site j is chosen randomly and a random amount of the Nth spin component is proposed for exchange such that $S_i^N + S_j^N$ remains constant. Then new spin components $(S_i^1, S_i^2, \ldots, S_i^{N-1})$ are proposed and the spin components $(S_j^1, S_j^2, \ldots, S_j^{N-1})$ are adjusted according to the spin normalization condition $|\boldsymbol{S}_i| = |\boldsymbol{S}_j| = 1$. The local change ΔE of the configurational energy is calculated according to (9). According to detailed balance the proposed update is accepted with probability $p(\beta \Delta E)$, where $\beta = 1/(k_B T)$. For our simulation we have chosen $p(x) = 1/(\exp(x) + 1)$. Note that all updates must be proposed such that the new spin at lattice site i is taken from the uniform distribution on the unit sphere in N dimensions.

The Wolff algorithm also works the standard way [9], except that *only* the first $N-1$ components of the spins are used for the cluster growth, i.e., (7) and (9) are treated as planar ferromagnets. This means that a cluster update never changes the Nth component of any spin so that the Wolff algorithm is nonergodic in this case. The cluster update is still a valid Monte-Carlo step in the sense that it fulfills detailed balance, however, in order to provide a valid Monte-Carlo algorithm it has to be used together with the Metropolis algorithm described above in a hybrid fashion. The use of Wolff updates allows us to take advantage of improved estimators [11] for magnetic quantities.

The overrelaxation part of the algorithm performs a microcanonical update of the configuration in the following way. The local configurational energy has the functional form of a scalar product of the spins, where according to (7) and (9) only the first $N-1$ components are involved. With respect to the sum of its nearest neighbor spins each spin has a transverse component in the $(S_i^1, S_i^2, \dots, S_i^{N-1})$ plane which does not enter the scalar product. The overrelaxation algorithm scans the lattice sequentially, determines this transverse component for each lattice site and flips its sign. This overrelaxation algorithm is similar to the one used in [10] and it quite efficiently decorrelates subsequent configurations over a wider range of temperatures around the critical point than the Wolff algorithm. However, overrelaxation can only be applied for $N \geq 3$. In the following only the cases $N = 2$ (transverse Ising) and $N = 3$ (transverse XY) are considered.

In a typical hybrid Monte-Carlo step we use three Metropolis *(M)*, seven single cluster Wolff *(C)* updates for $N = 2$ and three Metropolis, five single cluster Wolff, and two overrelaxation updates *(O)* for $N = 3$ in the critical region of the models. The inividual updates are mixed automatically in the program so that the update sequences *(M C C M C C M C C C)* for $N = 2$ and *(M C C M O C M C C O)* for $N = 3$ are generated as one hybrid Monte-Carlo step. The shift register generator R1279 given by the recursion relation $X_n = X_{n-p} \oplus X_{n-q}$ for $(p, q) = (1279, 1063)$ is used as the random number generator. Generators like this are known to cause systematic errors in combination with the Wolff algorithm [12]. However, for lags (p, q) as large as the ones used here these errors will be far smaller than typical statistical errors. They are further reduced by the hybrid nature of our algorithm due to the presence of several Metropolis updates in one hybrid Monte-Carlo step [13].

The hybrid Monte-Carlo scheme described above is employed for lattice sizes L between $L = 20$ and $L = 80$. For each system size and temperature we perform at least 10 blocks of 10^3 hybrid steps for equilibration followed by 10^4 hybrid steps for measurements. Each measurement block yields an estimate for all static quantities of interest and from these we obtain our final estimates and estimates of their statistical error following standard procedures. At the critical point (see below) two or three times as many updates have been performed. The integrated autocorrelation time of the hybrid algorithm is determined by the autocorrelation function of the energy or, equivalently, the modulus of the order parameter, which yield the slowest modes for the Wolff algorithm. The autocorrelation times are generally rather short, at the critical point they range from about 5 hybrid Monte-Carlo steps for $L = 20$ to about 10 hybrid Monte-Carlo steps for $L = 80$. The values for the equilibration and measurement periods given above thus translate to roughly 100 and 1000 autocorrelation times, respectively. In order to obtain the best statistics for magnetic quantities a measurement is made after every hybrid Monte-Carlo step. All error bars quoted in the following correspond to one standard deviation. The simulations have been performed on the DEC alpha AXP workstation cluster at the Physics Department and on HP RISC8000 workstations at the Computer Center of the RWTH Aachen.

3 Ising Universality Class

For $N = 2$ (7) and (9) describe a classical Ising model in a (fixed) transverse field or with fixed transverse magnetization, respectively. In the following we will only consider the constrained model with the symmetric constraint $M = 0$ and with the constraint $m \equiv M/L^3 = 1/\sqrt{2}$. The symmetrically constrained model does not show Fisher renormalization [3] and we therefore use this case for tests of the algorithm and for the production of data representative of the Ising universality class in $d = 3$. The constraint $m = const. \neq 0$ breaks the $S_i^N \to -S_i^N$ symmetry of the model and Fisher renormatization should become visible within a certain temperature window around $T_c = T_c(m)$. The width of this window is of course a nonuniversal property of the model and in particular one expects this window to widen as m is increased. Due to the spin normalization condition m cannot exceed unity and one therefore also expects, that critical behavior becomes very difficult to resolve numerically if m is too close to its maximum value. Therefore, $m = 1/\sqrt{2}$ is chosen as a compromise between good resolution in the critical regime and a prominent Fisher renormalization effect.

The critical temperatures $T_c(m = 0)$ and $T_c(m = 1/\sqrt{2})$ are determined from temperature scans of the Binder cumulant ratio according to standard procedures [14]. We obtain the following reduced critical coupling constants $K_c(m) \equiv J/k_B T_c(m)$:

$$K_c(0) = 0.41638 \pm 0.00005 \quad \text{and} \quad K_c(1/\sqrt{2}) = 0.6371 \pm 0.0001. \qquad (10)$$

The corresponding estimates for the Binder cumulant ratio obtained for $m = 0$ and $m = 1/\sqrt{2}$ agree with previous estimates obtained for the Ising universality class within two standard deviations [15], where for the latter choice of m Wegner corrections to scaling are considerable and must be subtracted in order to obtain a reliable estimate. In order to obtain an estimate for the exponent ν which enters the finite-size scaling argument according to (6) the cumulant

$$X \equiv \frac{\partial}{\partial T} \ln\langle\phi^2\rangle = \frac{1}{k_B T^2}\left(\frac{\langle\phi^2\mathcal{H}_M\rangle}{\langle\phi^2\rangle} - \langle\mathcal{H}_M\rangle\right) \tag{11}$$

has been measured, where $\phi = L^{-3}\sum_i S_i^1$ is the order parameter. At the critical temperature $T_c(m)$ the scaling behavior $X \sim x/t$ is expected (see (6)). Corresponding numerical results for $m = 0$ and $m = 1/\sqrt{2}$ are displayed in Fig.1 on a double logarithmic scale. The data are compatible with simple

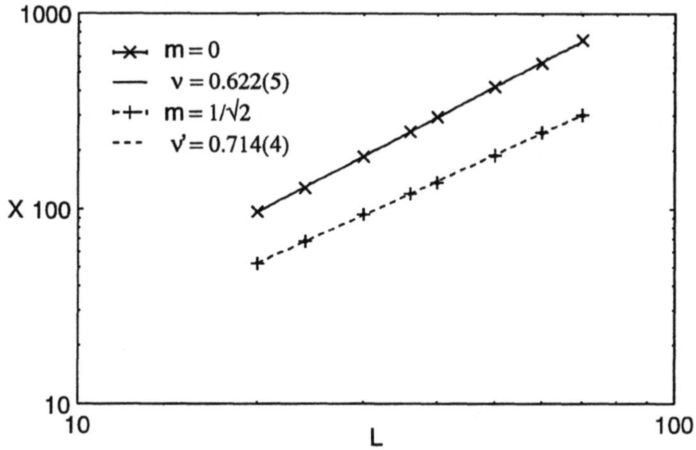

Fig. 1. Cumulant X at the critical point for $m = 0$ ($\times$) and $m = 1/\sqrt{2}$ (+). The solid and dashed lines display power law fits to the data for $30 \leq L \leq 70$ for $m = 0$ and $m = 1/\sqrt{2}$, respectively

power laws, where the exponents $\nu = 0.622 \pm 0.005$ ($m = 0$) and $\nu' = 0.714 \pm 0.004$ ($m = 1/\sqrt{2}$) have been obtained. Compared to the best currently known estimate $\nu \simeq 0.630$ [4] the above estimate is too small and only agrees with the theoretical value within two standard deviations. A more thorough analysis shows that the discrepancy can be explained by a mismatch of the order 5×10^{-5} between the actual critical temperature and the estimate used here (see (10)), which on the other hand is of the same magnitude as the statistical error of $K_c(0)$. The agreement between the above estimate for $\nu' = \nu/(1-\alpha)$ and the theoretical value $\nu' \simeq 0.708$ [4] is better, however, it may again be affected by a mismatch between the actual value of $T_c(1/\sqrt{2})$ and the estimate used here. If the literature values for ν and α are substituted in (6), where a and A are used as fit parameters, $a/A \simeq 0.1$ is obtained which is small enough to be ignored in the scaling analysis (see below).

The finite-size scaling analysis has been performed for several thermodynamic quantities, in particular, the average modulus of the order parameter $\langle|\phi|\rangle$, the susceptibilities

$$\chi_+ \equiv \frac{L^3}{k_B T}\langle\phi^2\rangle, \quad \chi_- \equiv \frac{L^3}{k_B T}\left(\langle\phi^2\rangle - \langle|\phi|\rangle^2\right), \tag{12}$$

and the specific heat C. Data will only be shown for $\langle|\phi|\rangle$, χ_-, and C, because the finite-size scaling functions for $\langle|\phi|\rangle$ and χ_+ are very similar. According to finite-size scaling theory it must be possible to callapse the data for all m onto one and the same curve, where two nonuniversal scaling factors are required for each quantity. One scaling factor adjusts the magnitude of the scaling argument x (see (6)), the other adusts the absolute normalization of the quantity. Note that the former saling factor must be the same for *all* quantities. For $m = 0$ the scaling argument $x = tL^{1/\nu}$ is used, whereas for $m = 1/\sqrt{2}$ the choice $x = tL^{(1-\alpha)/\nu}/A$ has been made, where $A \simeq 1.1$ and the coefficient a in (6) has been neglected. The exponents ν and α are taken from the literature [4]. The scaling plot of $\langle|\phi|\rangle$ is shown in Fig.2, where the

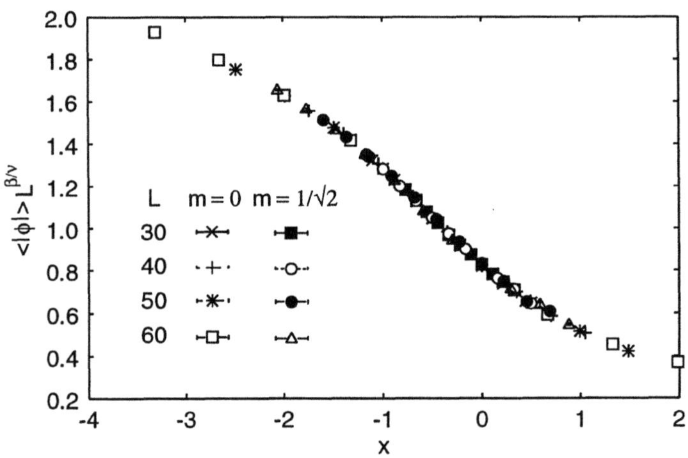

Fig. 2. Scaling plot of $\langle|\phi|\rangle$ for $L = 30$, 40, 50, and 60 for $m = 0$ and $m = 1/\sqrt{2}$. The reduced temperature t has been varied between -0.007 and 0.003. Statistical errors are much smaller than the symbol sizes

abolute normalization of the data for $m = 1/\sqrt{2}$ can be adjusted to the $m = 0$ data by a scale factor of ~ 0.7 as one would expect from simple mean field arguments. As shown in Fig.2, the espected data collapse can be reproduced rather well, where the literature value for the exponent $\beta/\nu = 0.5168$ [4] has been used. The same holds for the susceptibility χ_- which is displayed in Fig.3, where $\gamma/\nu = 2 - \eta = 1.967$ is also taken form [4]. The absolute magnitudes of $\chi_-/L^{\gamma/\nu}$ for $m = 0$ and $m = 1/\sqrt{2}$ are different by a factor of about 0.5 which is in accordance with simple mean field arguments. Note that the scaling function of χ_- has a maximum for $x < 0$ [16].

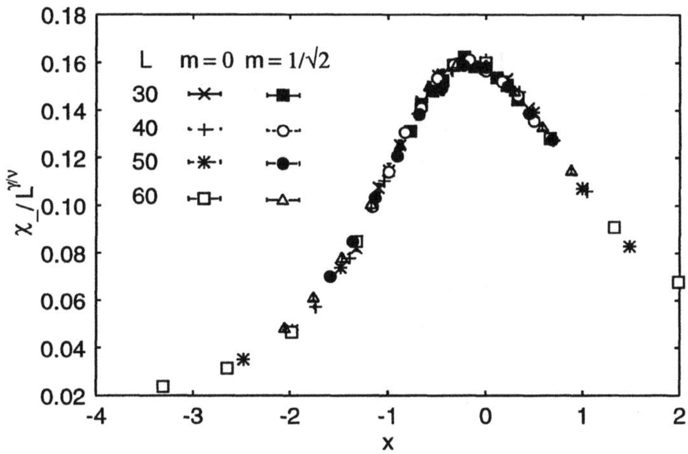

Fig. 3. Scaling plot of χ_- for $L = 30$, 40, 50, and 60 for $m = 0$ and $m = 1/\sqrt{2}$. The reduced temperature t has been varied between -0.007 and 0.003. Statistical errors are much smaller than the symbol sizes

The specific heat C requires a somewhat different treatment due to the fact that unlike the other quantities presented so far the specific heat requires an *additive* renormalization within renormalized field theory [16]. For the data analysis this means that scaling can only be obtained after a suitable subtraction is applied to the specific heat. One option to obtain scaling is to subtract the bulk specific heat $C_b^0(t)$ at a reference reduced temperature $t = t_0$ which are given by

$$C_b^0(t) = \frac{A_\pm}{\alpha}|t|^{-\alpha} + B \quad \text{and} \quad t_0 = (L/\xi_\pm^0)^{-1/\nu}, \tag{13}$$

where $A_\pm$ and B are nonuniversal constants and $\xi_\pm^0$ is the amplitude of the correlation length. The index $\pm$ refers to temperatures above or below $T_c(m)$, respectively. The reference reduced temperature chosen in (13) is positive and therefore only A_+ and ξ_+^0 are needed. Specifically, the choice $\nu = 0.630$, $\alpha = 0.109$ [4], $A_+ = 0.1552$, $B = -1.697$, and $\xi_+^0 = 0.495$ [16] guarantee scaling of the *relative* specific heat $\Delta C^0 \equiv C - C_b^0(t_0)$ for the Ising model in $d = 3$. Note that ξ_+^0 is measured in units of the lattice constant. For the data to be analyzed here the subtraction defined by (13) is only valid for the case $m = 0$, where ΔC^0 scales as $L^{\alpha/\nu}$. For $m = 1/\sqrt{2}$ Fisher renormalization according to (1) must be applied to $C_b^0(t)$ in order to obtain the correct form $C_b^m(t)$ of the subtraction. The result is

$$C_b^m(t) = \frac{A'_+}{\alpha}|t|^{\alpha/(1-\alpha)} + B' \quad \text{and} \quad t_0 = (L/\xi_\pm^0)^{(\alpha-1)/\nu}, \tag{14}$$

where $A'_+ = -0.1728$, $B' = 1.598$, and $\xi_+^0 = 0.495$ is not changed. The scaling factor $L^{\alpha/\nu}$, which usually governs the finite-size scaling of the specific heat, is cancelled here, i.e., one expects data collapse for $\Delta C^0/L^{\alpha/\nu}$ and $\Delta C^m \equiv C - C_b^m(t_0)$ up to an overall scale factor of about 0.5. The result of the data analysis is shown in Fig.4. The data collapse reasonably well onto

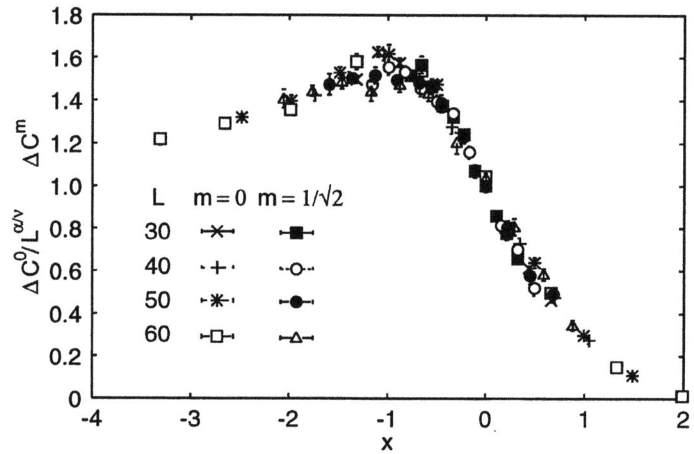

Fig. 4. Relative specific heats $\Delta C^0/L^{\alpha/\nu}$ for $m = 0$ and ΔC^m for $m = 1/\sqrt{2}$ for $L = 30$, 40, 50, and 60. The reduced temperature t has been varied between -0.007 and 0.003

a single curve except near the maximum of the scaling function, where also the scatter of the individual data is substantial due to a few bad samples for $L = 50$ and 60. However, there are also systematic deviations from scaling in the data, because the maximum in the scaling functions for the $m = 0$ data is more pronounced than in the $m = 1/\sqrt{2}$ data. These deviations could be due to enhanced Wegner corrections to scaling for $m = 1/\sqrt{2}$ as compared to $m = 0$.

4 XY Universality Class

For $N = 3$ (7) and (9) describe a classical XY model in a (fixed) transverse field or with fixed transverse magnetization, respectively. As for the case $N = 2$ we will only consider the constrained model with the symmetric constraint $m = 0$ and with the constraint $m = 1/\sqrt{2}$ in the following. The symmetrically constrained model is again used for algorithmic tests and data production for the XY universality class. The nonsymmetric constrained XY model does not show Fisher renormalization, however, according to (6) very slowly decaying corrections to the asymptotic critical behavior are expected, which will be discussed in the following. First, the cumulant X defined by (11) is evaluated at the critical point, which is given by the reduced coupling constants

$$K_c(0) = 0.6444 \pm 0.0001 \quad \text{and} \quad K_c(1/\sqrt{2}) = 1.1126 \pm 0.0003, \qquad (15)$$

respectively. The result for X is displayed in Fig.5. For $m = 0$ the data can

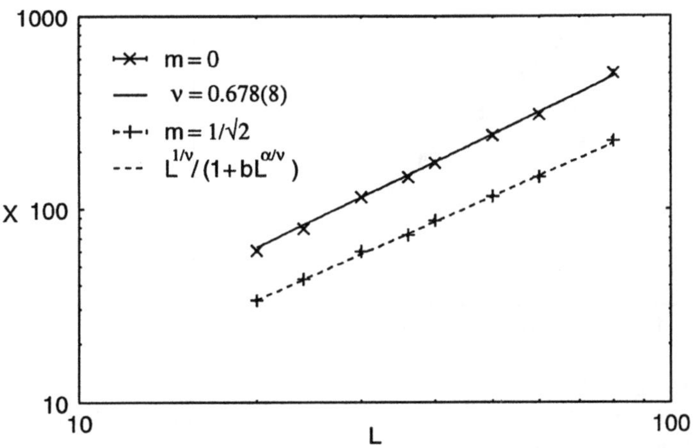

Fig. 5. Cumulant X at the critical point for $m = 0$ (x) and $m = 1/\sqrt{2}$ (+). The solid and dashed lines display fits to the data for $30 \leq L \leq 80$ for $m = 0$ and $m = 1/\sqrt{2}$, respectively (see main text)

be fitted by a power law $\sim L^{1/\nu}$, where $\nu = 0.678 \pm 0.008$ is obtained which agrees with the best current estimate $\nu = 0.671$ [4]. For $m = 1/\sqrt{2}$ the data can also be fitted by a power law, however, the resulting exponent ν only has the meaning of an effective exponent which does not fit into the XY universality class. As shown in Fig.5 the expression x/t according to (6) also yields a very good representation of the data where the parameter b in Fig.5 is given by $b = A/a = -0.941$. The exponents used in the fit (XY universality class) are taken from [4]. The value of the Binder cumulant found here agrees with results reported in the literature for the standard (plane rotator) XY model [17], however, Wegner corrections to scaling become quite substantial for $m = 1/\sqrt{2}$.

In the following scaling analysis the finite-size scaling argument for the case $m = 0$ takes its standard form $x = tL^{1/\nu}$ and for $m = 1/\sqrt{2}$ the combination $x = tL^{1/\nu}/(1 + bL^{\alpha/\nu})$ for $b = -0.941$ takes care of the slowly decaying correction terms to the asymptotic critical behavior caused by the very small and negative value of α in the XY universality class. As in Sect. 3 we consider $\langle|\phi|\rangle$, χ_+, χ_-, and the specific heat C in the scaling analysis. The scaling functions for $\langle|\phi|\rangle$ and χ_+ again look very similar so that we do not reproduce scaling plots for χ_+ here. The result for $\langle|\phi|\rangle$ is shown in Fig.6, where the order parameter $\phi \equiv L^{-3}\sum_i(S_i^1, S_i^2)$ has two components here. The data collapse very well onto a single curve. Note that the absolute

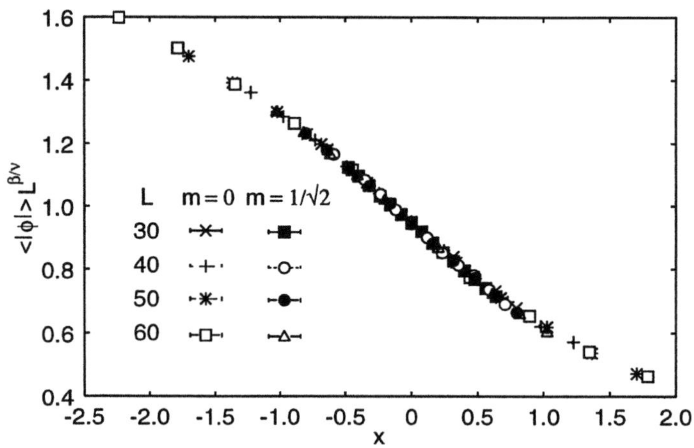

Fig. 6. Scaling plot of $\langle|\phi|\rangle$ for $L = 30$, 40, 50, and 60 for $m = 0$ and $m = 1/\sqrt{2}$. The reduced temperature t has been varied between -0.005 and 0.005. Statistical errors are much smaller than the symbol sizes

magnitudes of $\langle|\phi|\rangle$ for $m = 0$ and $m = 1/\sqrt{2}$ are again related by a factor of ~ 0.7 as suggested by mean-field arguments. The values for the critical exponents $\nu = 0.671$ and $\beta = 0.347$ are taken from the literature [4]. The susceptibility χ_- can be treated essentially as described in Sect.3, where the exponent $\gamma/\nu = 2 - \eta = 1.965$ is taken from [4]. The result of the scaling analysis is displayed in Fig.7. The data do not collapse as well as in Fig.3. Especially near the maximum of the scaling function the scatter of the data is substantially larger than in Fig.3. Slight systematic deviations from scaling for $m = 1/\sqrt{2}$ are observed which may again be due to enhanced Wegner corrections to scaling as compared to $m = 0$. Note that contrary to Fig.3 the scaling function of χ_- has a maximum for $x > 0$ in the XY universality class.

The specific heat C of the XY model also requires a subtraction before scaling is obtained [18]. The subtraction $C_b^0(t_0)$ is again used in the form given by (13), where $A_+ = 0.42$, $\alpha = -0.013$, $B = -A_+/\alpha$, and $\xi_+^0 = 1.0$ are used here which differ somewhat from the choices made for the standard XY model in [18]. It turns out, that the quality of the data collapse for the relative

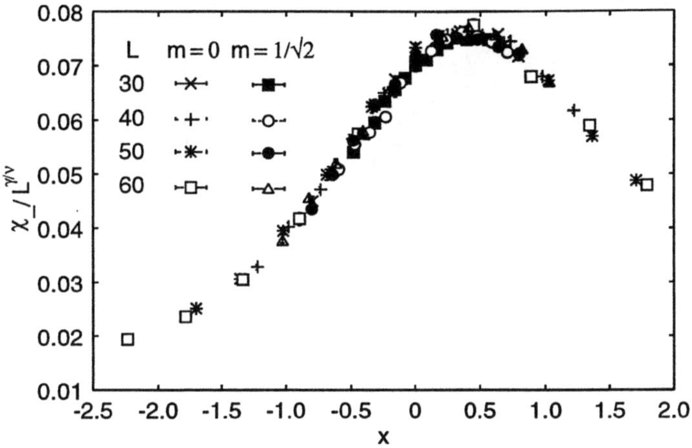

Fig. 7. Scaling plot of χ_- for $L = 30$, 40, 50, and 60 for $m = 0$ and $m = 1/\sqrt{2}$. The reduced temperature t has been varied between -0.005 and 0.005. Statistical errors are much smaller than the symbol sizes

specific heat $\Delta C^0 = C - C_b^0(t_0)$ for $m = 0$ is rather insensitive to the choice of ξ_+^0. The form of the subtraction $C_b^m(t_0)$ for $m = 1/\sqrt{2}$ requires a little analysis in order to include the slowly varying corrections to the asymptotic behavior coming from (1). One obtains the approximate form

$$C_b^m(t_0) = \frac{A_+/\alpha}{1+ct_0^{-\alpha}}\left[\left(\frac{t_0}{1+ct_0^{-\alpha}}\right)^{-\alpha} - 1\right], \quad t_0 = L^{-1/\nu}(1+bL^{\alpha/\nu}), \tag{16}$$

where $b = -0.941$ (see Fig.5), $A_+ = 0.42$ as before, and $c \simeq 2.0$ for optimal data collapse. The resulting scaling plot of ΔC^0 and $\Delta C^m \equiv C - C_b^m(t_0)$ is displayed in Fig.8. The overall shape of the scaling function is similar to the one shown in Fig.4. However, the scatter of the data near the maximum is so strong, that data collapse cannot be obtained in this region. In part this deficiency in the data may be due to the presence of 'bad' samples, e.g, for $L = 50$ and $m = 0$ at $t = -0.003$ and $t = -0.002$ and for $L = 60$ and $m = 1/\sqrt{2}$ at $t = -0.005$ and $t = -0.003$. Apart from that deviations from scaling as in Fig.4 may be present which are due to an enhancement of Wegner corrections to scaling for $m = 1/\sqrt{2}$ as compared to the case $m = 0$. However, Figs.6 - 8 confirm, that the choice of the scaling variable x given by (6) captures the slowly decaying correction term inside the scaling argument in an appropiate way and that furthermore the finite-size scaling behavior of constrained models in the Ising and the XY universality class can be treated on the same footing.

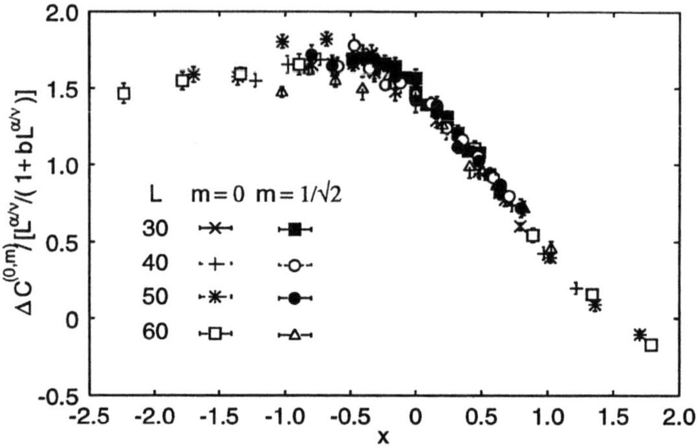

Fig. 8. Relative specific heats $\Delta C^{(0,m)}/\left[L^{\alpha/\nu}/(1+bL^{\alpha/\nu})\right]$ with $b=0$ for $m=0$ and $b=-0.941$ for $m=1/\sqrt{2}$ for $L=30$, 40, 50, and 60. The reduced temperature t has been varied between -0.005 and 0.005

5 Summary and Conclusions

The influence of constraints on the critical finite-size scaling behavior of Ising and XY models has been investigated by Monte-Carlo simulations of $O(N-1)$ planar ferromagnets with fixed transverse magnetization. The theoretical idea that only the form of the scaling argument is modified, whereas the shape of the universal scaling functions remains unchanged is verified within the statistical uncertainty of the data for the modulus of the order parameter, the susceptibilites χ_+ (not shown) and χ_-, and the specific heat. The form of the scaling argument used here allows to deal with critical finite-size effects in constrained Ising and XY models on the same footing, where constrained Heisenberg models can be included as well. Within the Ising universality class the finite-size behavior is consistent with the Fisher renormalization of critical exponents. In the XY universality class slowly decaying corrections to the asymptotic critical behavior are generated which are captured systematically by the analytic form of the scaling argument. The treatment of these corrections within the XY universality class may serve as a paradigm for the finite-size scaling analysis of dynamic quantities, where the constraints imposed here reappear as conserved quantities which are statically or dynamically coupled to the order parameter. These corrections may also be important for the interpretation of spin dynamics data for planar ferromagnets, where the energy of the system is conserved during the simulated time evolution of spin models similar to the ones investigated here.

Acknowledgments

The author gratefully acknowledges many helpful discussions with V. Dohm and financial support of this work through the Heisenberg program of the Deutsche Forschungsgemeinschaft.

References

1. Amit D. J. (1978) Field Theory, the Renormalization Group, and Critical Phenomena, McGraw-Hill, New York
2. Parisi G. (1988) Statistical Field Theory, Addison-Wesley, Wokingham
3. Fisher M. E. (1968) Phys. Rev. **176**, 257
4. Le Guillou J. C., Zinn-Justin J. (1985) J. Phys. (France) Lett. **46**, L137; Guida R., Zinn-Justin J. (1998) J. Phys. A **31**, 8103
5. Lipa J. A., Swanson D. R., Nissen J. A., Chui T. C. P., Israelsson U. E. (1996) Phys, Rev. Lett. **76**, 944
6. Fisher M. E. (1971) in: Green M. S. (Ed.) Proceedings of the 1970 Enrico Fermi School of Physics, Varenna, Italy, Course No. LI Academic, New, York, p.1; Fisher M. E., Barber M. N. (1972) Phys. Rev. Lett. **28**, 1516
7. Barber M. N. (1983) in: Domb C., Lebowitz J. L. (Eds.) Phase Transitions and Critical Phenomena, Vol. 8, Academic, New York, p. 145; Privman V. (1990) in: Privman V. (Ed.) Finite Size Scaling and Numerical Simulation of Statistical Systems, World Scientific, Singapore
8. Dohm V. (1998) private communication
9. Wolff U. (1989) Phys. Rev. Lett. **62**, 361
10. Chen K., Landau D. P. (1994) Phys. Rev. B **49**, 3266
11. Hasenbusch M. (1990) Nucl. Phys. **B333**, 581
12. Ferrenberg A. M., Landau D. P., Wong Y. J. (1992) Phys. Rev. Lett. **69**, 3382; Shchur L. N., Blöthe H. W. J. (1997) Phys. Rev. E **55**, R4905
13. Ferrenberg A. M., Landau D. P. unpublished
14. Chen K., Ferrenberg A. M., Landau D. P. (1993) Phys. Rev. B **48**, 3249
15. Blöte H. W., Luijten E., Heringa J. R. (1995) J. Phys. A **28**, 6289
16. Esser A., Dohm V., Chen X. S. (1995) Physica A **222**, 355; Chen X. S., Dohm V., Talapov A. L. (1996) Physica A **232**, 375
17. Gottlob A. P., Hasenbusch M. (1993) Physica A **201**, 593
18. Chen X. S., Dohm V., Esser, A. (1995) J. Phys. I France **5**, 205; Schultka N., Manousakis E. (1995) Phys. Rev. B **52**, 7528

Monte Carlo Simulation of Spin Models with Long-Range Interactions

E. Luijten

Max-Planck-Institut für Polymerforschung, D-55021 Mainz, Germany
Institut für Physik, Johannes Gutenberg-Universität, D-55099 Mainz, Germany

Abstract. An efficient Monte Carlo algorithm for the simulation of spin models with long-range interactions is discussed. Its central feature is that the number of operations required to flip a spin is independent of the number of interactions between this spin and the other spins in the system. In addition, critical slowing down is strongly suppressed. In order to illustrate the range of applicability of the algorithm, two specific examples are presented. First, some aspects of the Kosterlitz–Thouless transition in the one-dimensional Ising chain with inverse-square interactions are calculated. Secondly, the crossover from Ising-like to classical critical behavior in two-dimensional systems is studied for several different interaction profiles.

1 Introduction

For a long time, systems with long-range interactions have posed a great challenge to the application of Monte Carlo (MC) methods. Due to the large number of interactions that have to be taken into account, simulations are very time consuming and hence restricted to relatively small system sizes. Alternatively, the interaction is truncated beyond a certain distance, which may introduce severe errors in the calculation. Thus, it is desirable to have an algorithm at one's disposal that allows the efficient simulation of systems with long-range interactions without making any approximation except for the inherent statistical errors. A few years ago, an algorithm fulfilling these requirements was indeed introduced for the simulation of spin models [1]. It is based on the cluster method introduced by Swendsen and Wang [2] and has the crucial property that the number of operations required to add a single spin to a cluster is *independent* of the number of interactions of this spin with other spins in the system. Thus, the numerical effort to simulate ferromagnetic long-range models becomes comparable to that for short-range models. Close to the critical point the application of the algorithm becomes particularly profitable, since there one also benefits from the inherent reduction in critical slowing down in cluster algorithms. It is the purpose of the present paper to discuss this algorithm in some more detail and to illustrate its power by means of some selected applications.

Springer Proceedings in Physics, Volume 85
Computer Simulation Studies in Condensed-Matter Physics XII
Eds.: D.P. Landau, S.P. Lewis and H.-B. Schüttler

2 Description of the Algorithm

2.1 Conventional Wolff Cluster Algorithm

To set the stage, we first briefly discuss the standard cluster algorithm for a system with short-range interactions. For simplicity, we will employ a formulation in terms of the Wolff cluster algorithm [3] rather than that of Swendsen and Wang (SW). It should be noted, however, that the modifications required for the simulation of long-range interactions pertain to the cluster-formation process only, and hence can equally well be combined with the SW variant. Since a detailed derivation of the Wolff algorithm can be found in numerous references, we restrict ourselves here to a schematic outline. Consider a d-dimensional lattice structure, where on each lattice site i a spin s_i is placed, which can take the values ± 1 and which has a ferromagnetic coupling K with its nearest neighbors. Thus, the system is described by the following Hamiltonian,

$$\beta\mathcal{H} = -K \sum_{i<j} s_i s_j \ , \tag{1}$$

where the sum runs over all pairs of nearest neighbors. At each Monte Carlo step, a cluster of spins with identical sign is flipped. The nonlocal character of this operation leads to a rapid decay of correlations. The crucial step is the identification of a cluster:

1. Randomly choose a spin s_i from the lattice.
2. Consider each spin that interacts with spin s_i. It is added to the cluster with a probability $p = 1 - \exp(-2K)$, provided that it has the same sign as s_i.
3. Repeat step 2 in turn for each spin that is newly added to the cluster, where one now considers all spins that interact with this spin, rather than with s_i. This is iterated until all neighbors of all spins in the cluster have been considered for inclusion.

This cluster-building process is based on the Kasteleyn–Fortuin mapping [4,5] of the Potts model on a bond-percolation model. This mapping allows a generalization of the outlined procedure to systems with different interaction strengths K_j. In the second step, a spin is simply added with a probability $p_j = 1 - \exp(-2K_j)$ if the interaction strength with the spin s_i is equal to K_j. As a side note, it is remarked that this implies that a cluster now not necessarily can be visualized as a contiguous collection of spins.

Although one can thus easily devise a cluster algorithm for systems with long-range interactions, its efficiency rapidly decreases with an increasing number of interactions. Indeed, the typical value for the coupling K in such a system will be relatively small and hence each single spin that is considered for inclusion in the cluster will be added only with a small probability. So, a lot of operations are required to add a single spin to the cluster.

2.2 Efficient Construction of Clusters

In order to illustrate our approach, we take the example of a one-dimensional chain with a spin–spin interaction that decays as a power-law, $K_{ij} = fr_{ij}^{-1-\sigma}$. The cluster-building process starts with a spin on a randomly chosen site i and all other spins in the system are added to the cluster with a probability $p(s_i, s_j) = \delta_{s_i s_j} p_{ij}$, where $p_{ij} = 1 - \exp[-2f|i-j|^{-(1+\sigma)}]$. For each spin that is actually added to the cluster, its address is also placed on the *stack*. When all spins interacting with the first one have been considered, we read a new spin from the stack and repeat the process, until the stack is empty. The spin from which we are currently adding spins will be called the *current spin*. In order to avoid considering each single spin for inclusion in the cluster, we split up the probability $p(s_i, s_j)$ into two parts, namely the Kronecker delta testing whether the spins s_i and s_j have the same sign and the "provisional" probability p_{ij}. This enables us to define the concept of the *cumulative probability* $C(j)$, from which we can read off which spin is the next one to be provisionally added to the cluster,

$$C(j) \equiv \sum_{n=1}^{j} P(n) \tag{2}$$

with

$$P(n) = \left[\prod_{m=1}^{n-1} (1 - p_m) \right] p_n \ . \tag{3}$$

$p_j \equiv 1 - \exp(-2K_j)$ is an abbreviation for p_{0j} (and $K_j \equiv K_{0j}$), i.e., we define the origin at the position of the current spin. $P(n)$ is the probability that in the first step $n-1$ spins are skipped and the nth spin is provisionally added. Now the next spin j that is provisionally added is determined by a (pseudo)random number $g \in [0,1]$: $j-1$ spins are skipped if $C(j-1) \leq g < C(j)$. If the jth spin has indeed the same sign as the current spin then s_j is added to the cluster. Subsequently we skip again a number of spins before another spin at a distance $k > j$ is provisionally added. Due to the requirement $k > j$ we must shift the function P,

$$P_j(k) = \left[\prod_{m=j+1}^{k-1} (1 - p_m) \right] p_k \ , \tag{4}$$

and Eq. (3) is simply a special case of Eq. (4). The appropriate cumulative probability is now given by a generalization of Eq. (2),

$$C_j(k) = \sum_{n=j+1}^{k} P_j(n) \ . \tag{5}$$

By using the specific form of the probability p_{ij} one finds that this reduces to

$$C_j(k) = 1 - \exp\left(-2 \sum_{n=j+1}^{k} K_n\right) , \tag{6}$$

i.e., the probability that the next spin that will be added lies at a distance in the range $[j+1, k]$ is given by an expression that has the same form as the original probability, in which the coupling constant is replaced by the sum of all the couplings with the spins in this range.

We now consider two possibilities of calculating the distance k from a given $C_j(k)$. A straightforward approach is the construction of a look-up table. This means that we explicitly carry out the sum in (6) for a large number of distances k, up to a certain cutoff M, and store the results in a table. Then, after drawing a random number, we can derive the corresponding spin distance from this table. In principle we need for each value of j another look-up table containing the $C_j(k)$. This is hardly feasible and fortunately not necessary, as follows from a comparison of Eqs. (2) and (5). Namely (assume $k > j$),

$$\begin{aligned} C(k) &= C_0(k) = C(j) + \left[\prod_{i=1}^{j}(1-p_i)\right] C_j(k) \\ &= C(j) + [1 - C(j)]\, C_j(k) \end{aligned} \tag{7}$$

or $C_j(k) = [C(k) - C(j)]/[1 - C(j)]$. So we can calculate $C_j(k)$ directly from $C(k)$. In practice one realizes this by using the distance j of the previous spin that was provisionally added to rescale the (new) random number g to $g' \in [C(j), 1]$; $g' = C(j) + [1 - C(j)]g$. Since we only consider ferromagnetic interactions, $\lim_{j\to\infty} C(j)$ exists and is smaller than 1, cf. Eq. (6).

This method is very fast, since we have to calculate all cumulative probabilities only once, but it has two major drawbacks. First, we can accommodate only a limited number of spin distances in our look-up table and must therefore devise some approximation scheme to handle the tail of the long-range interaction, which is essential for the critical behavior in the case of slowly decaying interactions (small σ). This issue is addressed below. Secondly, this method is impractical in more than one dimension, as the number of distances for which the cumulative probability has to be calculated rapidly increases with the dimensionality of the system (for a fixed cutoff).

For interactions that can be explicitly summed there exists a powerful alternative. In those cases, Eq. (6) can be solved for k, yielding an expression for the spin distance in terms of $C_j(k)$, i.e., in terms of the random number g. Especially for the study of critical phenomena this is often a feasible approach, since in many cases one can modify an interaction such it can be explicitly integrated by only adding irrelevant terms that do not affect the universal critical properties. As an example, we consider again the power-law interaction $K_{ij} = f|i-j|^{-(1+\sigma)}$. The corresponding sum appearing in the right-hand side of (6) is (for $j = 0$) the truncated Riemann zeta function, which cannot be expressed in closed form. However, upon approximation of this sum by the integral

$$\sum_{n=j}^{k} K_n = \sum_{n=j}^{k} \frac{f}{n^{1+\sigma}} \approx f \int_{j-\frac{1}{2}}^{k+\frac{1}{2}} dx\, x^{-(1+\sigma)} \tag{8}$$

one still has an exact Monte Carlo scheme, but for an interaction

$$K(|i-j|) = f \int_{|i-j|-\frac{1}{2}}^{|i-j|+\frac{1}{2}} dx\, x^{-(1+\sigma)} . \tag{9}$$

Both interactions belong to the same universality class, so that, e.g., for the determination of critical exponents one is free to make the most convenient choice. Of course, all nonuniversal quantities, such as the critical temperature, will have different values. For the modified interaction, Eq. (6) reduces to

$$C_j(k) = 1 - \exp\left[-\frac{2f}{\sigma}\left(\frac{1}{j^\sigma} - \frac{1}{k^\sigma}\right)\right] , \tag{10}$$

where it should be noted that this is only the probability for adding a spin that lies in a single direction. We have not written $j + \frac{1}{2}$ and $k + \frac{1}{2}$ instead of j and k, because we prefer to use continuous coordinates, which only in the last stage are rounded to a lattice site. Equating $C_j(k)$ to the random number g we find

$$k = \left[j^{-\sigma} + \frac{\sigma}{2f}\ln(1-g)\right]^{-1/\sigma} . \tag{11}$$

Rescaling of the random number is no longer required: the lowest value, $g = 0$, leads to a provisionally added spin at the same distance as the previous one, $k = j$. If $g = C_j(\infty) = 1 - \exp[-(2f/\sigma)j^{-\sigma}]$ the next provisionally added spin lies at infinity and thus $g \in [C_j(\infty), 1]$ yields no spin at all. Once the coordinates of the next provisionally added spin have been calculated by rounding to the nearest integer coordinate, the periodic boundary conditions are applied to map this coordinate onto a lattice site. For the following provisionally added spin, j is set equal to k (*not* to the rounded distance!) and a new k is determined. If no spin has been added yet, j is set to $\frac{1}{2}$, the lowest possible distance. For higher dimensionalities, $d - 1$ additional random numbers are required to determine the direction in which the next spin is added.

The above-mentioned limited size M of the look-up table can now be coped with as well: beyond the spin distance M the sum in (6) is approximated by an integral. Thus, if the random number g lies in the interval $[C(M), C(\infty)]$, the spin distance k is determined from a modified version of (11), where the lower part of the integral is replaced by an explicit sum

$$k = \left[\left(M + \frac{1}{2}\right)^{-\sigma} + \sigma\left(\frac{1}{2f}\ln(1-g) + \sum_{n=1}^{M} \frac{1}{n^{1+\sigma}}\right)\right]^{-1/\sigma} . \tag{12}$$

2.3 Discussion

Finally, we briefly illustrate the efficiency of the algorithm by means of an example. If the probability p were equal for each spin pair, one out of p^{-1} spins would be added to the cluster, and it would take $\mathcal{O}(p^{-1}) \sim \mathcal{O}(L^0)$ operations per spin to update a configuration, compared to $\mathcal{O}(L^d)$ operations *per spin* for a Metropolis algorithm. Taking into account the decrease in critical slowing down, we see that the efficiency of this method is typically a factor $\mathcal{O}(L^{d+z})$ larger than the conventional Monte Carlo algorithm. In the case of a large but finite interaction range R, such as for the two-dimensional equivalent-neighbor model discussed in Sec. 3.2, one gains a factor $\mathcal{O}(R^d L^z)$, provided that critical slowing down is completely suppressed in the cluster algorithm. Far from the critical temperature, the factor L^z disappears, but one still has the advantage of a speed increase proportional to the number of interactions per spin.

3 Applications

3.1 The Inverse-Square Ising Chain

As a first application, we consider the Ising chain introduced before, with spin–spin interactions that decay as $fr_{ij}^{-1-\sigma}$. Unlike the Ising chain with short-range interactions, in which no long-range order is possible for nonzero temperatures, this system exhibits a remarkably rich behavior. The interested reader is referred to Ref. [6] for a review. In summary, the system exhibits mean-field-like critical behavior for $0 < \sigma \leq \frac{1}{2}$ and nonclassical critical behavior for $\frac{1}{2} < \sigma < 1$, with critical exponents that depend on σ. For $\sigma > 1$ the interactions are essentially short-ranged. At $\sigma = 1$, the so-called inverse-square Ising model, we have a very interesting situation: the spin chain displays a phase transition which is the one-dimensional analog of the Kosterlitz–Thouless (KT) transition in the two-dimensional XY model. It has close connections to a variety of physical applications, such as the Kondo problem, quantum tunneling in a two-state system coupled to a dissipative environment, quark confinement, etc. Although the KT transition has received an enormous amount of attention over the past decades, there are still open questions. Two peculiar properties of this transition are: (1) at the critical temperature T_c the order parameter exhibits a singular behavior *superposed* on a jump like one finds for a first-order transition; (2) the correlation length ξ and the susceptibility χ diverge exponentially for $T \downarrow T_c$: $\xi = \xi_0 \exp[B_\xi/(T-T_c)^\nu]$ and $\chi = \chi_0 \exp[B_\chi/(T-T_c)^\nu]$, with $\nu = \frac{1}{2}$. These and related critical properties turn out to be very difficult to verify in numerical simulations of two-dimensional models, because the finite-size effects decay only logarithmically. The one-dimensional model, which now can be simulated with comparable efficiency, clearly offers a great advantage: rather than a linear system size of $\mathcal{O}(10^3)$, like for $d = 2$, one can reach system sizes

$L = \mathcal{O}(10^6)$ and thus approach the critical point much closer. Since a detailed description of the critical properties and their numerical determination is beyond the scope of this paper, we restrict ourselves here to a study of the quantity $\Psi \equiv Km^2$, where K is the coupling constant and m the magnetization density, and its application for the determination of T_c (or, equivalently, its inverse, the critical coupling K_c). It can be shown that Ψ is the analog of the spin-wave stiffness in the 2D XY model. Just like the latter quantity is predicted to have a universal jump $2/\pi$ at criticality, Ψ is expected to have a jump of size $\frac{1}{2}$ [7]. Indeed, in Figure 1, where $\Psi(K, L)$ is shown for system sizes up to $L = 400\,000$, one can already clearly observe how such a jump develops with increasing system size. The superposed square-root singularity,

$$\Psi(K, \infty) = \Psi(K_c, \infty) + C\sqrt{K - K_c} + \mathcal{O}(K - K_c) \,, \tag{13}$$

is shown in the inset.

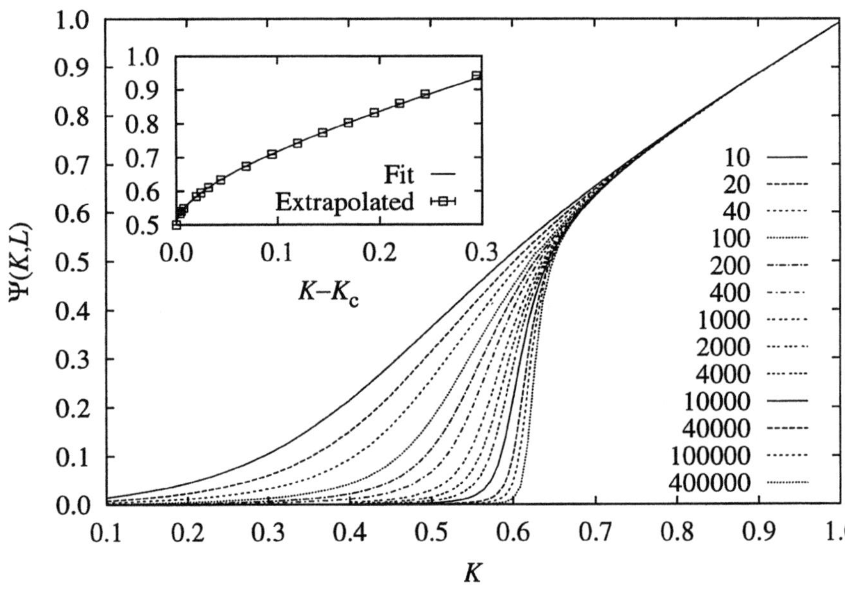

Fig. 1. The quantity $\Psi = Km^2$ as a function of the coupling K, for system sizes $10 \leq L \leq 400\,000$. The inset shows $\lim_{L\to\infty} \Psi(K, L)$ for $K > K_c$

We consider now three distinct ways to use this quantity for the determination of K_c, which for phase transitions in this universality class is notoriously difficult. First, one may use the predicted singular behavior of Ψ for $K > K_c$. To this end, the finite-size data at fixed couplings first must be extrapolated to the thermodynamic limit. By integrating the RG equations one finds [7] that $\Psi(K,L)$ obeys a finite-size expansion of the form

$$\Psi(K,L) = \Psi(K,\infty)\left\{1 + a_1 L^{-2[\bar{\Psi}-1]} + a_2 L^{-4[\bar{\Psi}-1]} + \cdots\right\} , \tag{14}$$

where $\bar{\Psi} = \Psi(K,\infty)/\Psi(K_c,\infty)$ and the ellipsis denotes higher-order terms. The resulting estimates for $\Psi(K,\infty)$ are then fitted to Eq. (13), in which $\Psi(K_c,L)$ is kept fixed at $\frac{1}{2}$. This yielded $K_c = 0.6552$ (2). Secondly, $\Psi(K,L)$ in the high-temperature regime $K < K_c$ may be used to estimate K_c. Indeed, within the finite-size regime one expects two types of corrections to scaling: corrections due to irrelevant fields, which decay as powers of $1/\ln L$, and temperature-dependent corrections which can be expanded in terms of L/ξ. A least-squares fit of the numerical data to an expansion of the form

$$\Psi(K,L) = \Psi(K_c,\infty) + a_1\frac{L}{\xi} + a_2\left(\frac{L}{\xi}\right)^2 + \cdots + \frac{b_1}{\ln L} + \frac{b_2}{(\ln L)^2} + \cdots \tag{15}$$

has yielded $\Psi(K_c,\infty) = 0.496$ (3) and $K_c = 0.6548$ (14). Fixing $\Psi(K_c,\infty)$ at the predicted value $\frac{1}{2}$ we found $K_c = 0.6555$ (4). Finally, a very straightforward but remarkably effective approach is to fit a set of finite-size data for $\Psi(K,L)$ at *fixed* coupling to an expression of the form of Eq. (15), in which all temperature-dependent terms have been omitted, i.e., one *assumes* that $K = K_c$. Although in principle such a fit should only work at the true critical coupling, it turns out that least-squares fits of a good quality can be obtained over a range of couplings. However, the resulting estimate of $\Psi(K,\infty)$ is a monotonously increasing function of K and K_c can be determined from the requirement that $\Psi(K_c,\infty) = \frac{1}{2}$, yielding $K_c = 0.65515$ (20).

Table 1. Some estimates (in chronological order) for the critical coupling K_c of the inverse-square Ising chain.

Ref.	K_c	Method
[8]	≈ 0.612	Exact summation + Padé approximants
[9]	≈ 0.635	RG
[10]	≈ 0.490	Extended mean-field approach
[11]	0.657 (3)	Series expansion
[12]	≥ 0.441	Extended mean-field approach
[13]	$\geq \frac{1}{2}$	Analytical
[14]	≈ 0.562	Coherent-anomaly method
[15]	0.5	Variational method
[16]	0.590 (5)	Cycle expansion
[17]	0.615	Transfer matrix
[18]	≥ 0.61128	Extended mean-field approach
[19]	$6/\pi^2 \approx 0.6079$	RG (conjectured to be exact)
This work	0.6552 (2)	MC

It is rewarding that the three different methods yield consistent estimates for the critical coupling. Table 1 summarizes several estimates for K_c, obtained by a variety of methods. Remarkably, most numerical results (of which only few carry an explicit estimate of the uncertainty) suggest a critical coupling around 0.61. Our estimate lies considerably higher and only agrees with the series-expansion result of Ref. [11]. All methods that rely on an extension of mean-field theory clearly suffer from the extremely slow convergence to the thermodynamic limit. The exact conjecture of Ref. [19] (which is, interestingly, precisely twice as large as the mean-field result $3/\pi^2$) has been refuted.

3.2 Crossover from Ising-Like to Classical Critical Behavior

In order to illustrate that also for interactions with a large but strictly finite range the presented algorithm offers great advantages, we consider the so-called "equivalent-neighbor" model. This is a simple generalization of the Ising model, in which each spin interacts equally strongly with all its neighbors within a distance R_m. In the limit $R_m \to \infty$ this model becomes equivalent to the exactly-solved mean-field model, whereas all systems with a *finite* interaction range belong to the Ising universality class. Since an exact solution for the latter case is lacking for $d = 3$, it is interesting to study the variation of critical properties as a function of R_m (cf. Ref. [20]). Another, experimentally very relevant, application of this model will be discussed here. As is well known, many thermodynamic properties show a characteristic power-law divergence upon approach of a critical point. These powers, or critical exponents, have universal values that only depend on a small number of *global* properties of the system under consideration. For example, binary mixtures, simple fluids, and uniaxial ferromagnets all exhibit the same set of critical exponents as the three-dimensional Ising model. However, the corresponding power-law behavior is only observed asymptotically close to the critical point, whereas at temperatures farther away from T_c (but still relatively close to it) one may observe classical or mean-field-like critical behavior. This *crossover* can be explained in terms of competing fixed points of a renormalization-group transformation and is in principle well understood. However, unlike the critical exponents, for which accurate results have been obtained from series expansions, renormalization-group calculations, experiments, and numerical calculations, the precise nature of the crossover from one universality class to another is still a point of discussion. In particular, it is an unsettled question to what extent this crossover is universal. There exist several field-theoretic calculations, but it has not yet been possible to verify their correctness by means of experiments. Measurements in the critical region are not only difficult, one also has to take great care to make the temperature distance to the critical point not too large, since one then would leave the critical region. As stated by the Ginzburg criterion, the crossover is a function of t/G, where $t = (T - T_c)/T_c$ is the reduced temperature and G a system-dependent parameter. Throughout the crossover region, t has to be kept small, but t/G has to be varied over several decades. The large extent of the crossover region also emphasizes its experimental relevance: many measurements of critical exponents are actually made within this region rather than asymptotically close to the critical point and hence a detailed knowledge of crossover functions is required for a proper interpretation of the data. Since the Ginzburg parameter G is a function of the interaction range, it is well possible to construct such crossover functions from data obtained by means of simulations of systems with different interaction ranges, where for each system t is varied over a limited range only. In view of the large coordination numbers $[\mathcal{O}(10^4)]$ that have to be reached within this approach, this is only feasible with an advanced algorithm.

We concentrate here on one specific crossover function, namely that for the susceptibility in a two-dimensional (2D) system, both below and above the critical temperature (see Refs. [21,1] for a more detailed discussion of this topic). In the 2D Ising model, the susceptibility χ diverges for $t \uparrow 0$ as $A_{\mathrm{I}}^{-}(-t)^{-7/4}$ and for $t \downarrow 0$ as $A_{\mathrm{I}}^{+}t^{-7/4}$, where the amplitudes $A_{\mathrm{I}}^{\pm}$ are known exactly. Mean-field theory, on the other hand, predicts a susceptibility that for $t \uparrow 0$ diverges as $1/(-2t)$ and for $t \downarrow 0$ as $1/t$. It is our aim to numerically determine the effective susceptibility exponent $\gamma_{\mathrm{eff}}^{\pm} = -d\ln\chi/d\ln t$, which is expected to interpolate smoothly between 7/4 and 1. We have carried out MC simulations for square lattices with a maximum linear size $L = 1000$ and interaction ranges up to $R_m^2 = 10000$ (coordination number $z =$

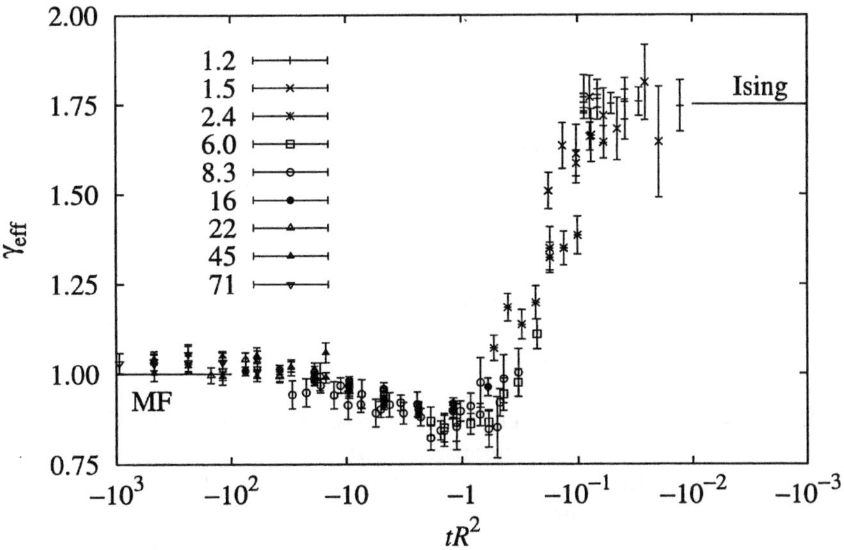

Fig. 2. The effective susceptibility exponent $\gamma_{\mathrm{eff}}^{-}$ for $T < T_c$ as a function of the crossover variable tR^2

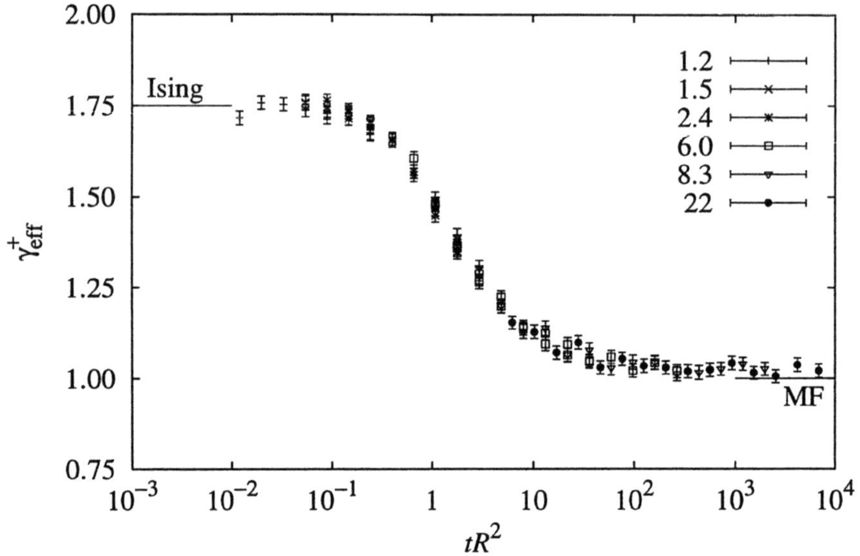

Fig. 3. The effective susceptibility exponent $\gamma^+_{\rm eff}$ for $T > T_c$ as a function of the crossover variable tR^2

31416). It has been shown that the Ginzburg parameter $G \propto R^{-2}$, where the effective interaction range $R^2 = z^{-1}\sum_{j\neq i}|\mathbf{r}_i - \mathbf{r}_j|^2$ $(|\mathbf{r}_i - \mathbf{r}_j| \leq R_m)$ has been introduced to avoid lattice effects. For $t < 0$ the susceptibility is calculated from the fluctuation relation $\chi' = L^d(\langle m^2\rangle - \langle |m|\rangle^2)/k_BT$, whereas $\chi = L^d\langle m^2\rangle/k_BT$ was used for $t > 0$. The exponent $\gamma^\pm_{\rm eff}$ is obtained by numerical differentiation and shown as a function of t/G in Figs. 2 and 3, respectively.

Several observations can be made concerning these graphs. In the first place, one notices that both functions smoothly interpolate between the Ising exponent and its classical counterpart. Also the expectation that the crossover takes several decades in the parameter t/G is confirmed. However, there is a striking difference between $\gamma^-_{\rm eff}$ and $\gamma^+_{\rm eff}$: whereas the latter exhibits a gradual decrease from 7/4 to 1 when the distance to the critical point is increased, the former drops much more rapidly and even goes through a minimum where $\gamma_{\rm eff} < 1$, before reaching its asymptotic value. Interestingly, the occurrence of such a nonmonotonic variation has been inferred from RG calculations for three-dimensional systems. The results presented here are the first to show that, at least in the low-temperature regime, this behavior can actually be observed in systems as simple as the two-dimensional Ising model with an extended interaction range.

The fact that the curves in Figs. 2 and 3 overlap for different values of R suggests that the crossover functions possess a certain degree of universality. However, to what extent one may expect such universality in a region vhere the correlation length no longer diverges is still an unanswered ques-ion. It has been suggested [23] that an additional length scale comes into ›lay here. In order to investigate this possibility, we have carried out sim-ılations of two-dimensional models that are very similar to those studied ıbove, except that the interaction profile has been modified [24]. Instead of ı constant interaction strength (block-shaped profile), a double-blocked po-ential was used, where each spin interacts with a strength K_1 with its z_1 ıeighbors within a distance $r \leq R_1$ (domain D_1) and with a strength K_2 vith its z_2 neighbors within a distance $R_1 < r \leq R_2$ (domain D_2). In order o create a strong asymmetry, a strength ratio $\alpha = K_1/K_2 = 16$ was cho-en. The outer interaction range was kept fixed at $R_2 = \sqrt{140}$, whereas the ›arameter R_1 was varied. This allowed us to realize two additional, qualita-ively different situations: both $R_1 = \sqrt{4}$ and $R_1 = \sqrt{93}$ lead to an effective nteraction range $R \approx \sqrt{50}$ [the previous definition for R is now generalized o $R^2 = (\alpha \sum_{i \in D_1} r_i^2 + \sum_{i \in D_2} r_i^2)/(\alpha z_1 + z_2)]$, but in the former case the nner domain contains only 12 out of 436 interaction neighbors, compared to !92 out of 436 in the latter case. This means that the integrated coupling atio $\alpha z_1/z_2$, which indicates the relative contribution of the two domains to he total integrated coupling, is 0.45 in the first case and 32 in the second :ase. Extensive Monte Carlo simulations have been carried out for these and ıeveral other double-blocked interaction profiles. After a determination of the :ritical properties, we have calculated the crossover curve for the susceptibil-ty in the region $t < 0$ for each of these profiles, in order to see whether the ıonmonotonicity in Fig. 2 is a peculiarity of the equivalent-neighbor model. The critical temperature, normalized by the critical temperature of the mean-ield model, indeed displays a clear nonuniversality: for $R_1^2 = 93$ the critical luctuations turn out to be stronger suppressed than for $R_1^2 = 4$, even though :he effective interaction range is almost identical. The data for the suscepti-›ility are shown in Fig. 4. Since the critical amplitude of the susceptibility raries as $R^{-3/2}$, a data collapse is obtained for χ'/R^2. One observes how all lata perfectly coincide with the crossover curve for the equivalent-neighbor nodel. The deviations from the curve at the right-hand side of the graph ıre caused by the fact that, sufficiently close to T_c, the diverging correlation ength is truncated by the finite system size. For the systems with $R_1^2 = 4$ ınd $R_1^2 = 93$ (both with $R_2^2 = 140$) this happens in the figure at different :emperatures, despite the fact that they have very similar values for the effec-:ive range R. The reason for this is that the data points pertain to different ›ystem sizes, viz. $L = 1000$ and $L = 300$, respectively. The inset shows that 'or the *same* system size ($L = 300$) the data points for both systems virtu-ılly coincide, even in the finite-size regime. The crossover function for the :onnected susceptibility appears to be insensitive to the introduction of an ıdditional length scale in the interaction profile. We hence view this graph ıs a strong indication that crossover functions possess a considerable degree ›f universality.

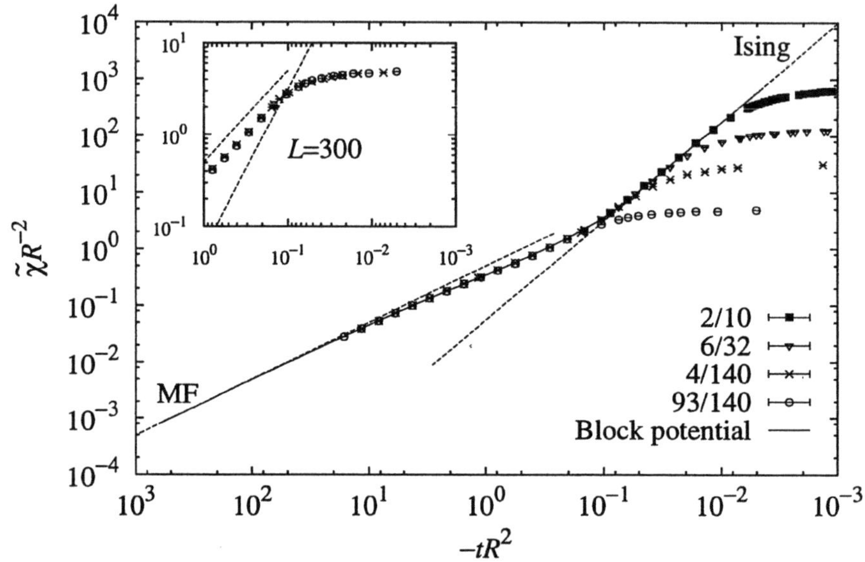

Fig. 4. The crossover function for the connected susceptibility below T_c, for the two-dimensional Ising model with a double-blocked interaction profile. $\tilde{\chi} = \chi'/C(R)$, where $C(R)$ is a range-dependent correction factor that accounts for the fact that the critical amplitude for small R deviates from the asymptotic range dependence. This only introduces a shift along the vertical axis. The solid curve indicates the crossover function for the block-shaped interaction profile and the dashed lines mark the mean-field ("MF") and the Ising asymptote. The numbers in the key refer to the values for R_1^2 and R_2^2 for each interaction profile

4 Outlook and Conclusion

We have discussed a cluster algorithm for spin models with long-range interactions that is several orders of magnitude more efficient than conventional algorithms. Its usefulness has been demonstrated for a system with power-law interactions as well as for systems with medium-range interactions. In both cases, physical results have been obtained that until now could not be generated within a reasonable amount of computing time. However, the scope of the algorithm goes far beyond what could be presented here. As far as purely ferromagnetic interactions are concerned, it can be efficiently applied to any interaction profile, including anisotropic ones. Also the order parameter must not necessarily be Ising-like: the extension to general $O(n)$ models [1] is similar to that for the original Wolff algorithm and the first application to a Potts model has already been published [25].

The generalization to a situation in which additional antiferromagnetic interactions are present is, in principle, also possible. Whereas competing interactions will move the system away from the percolation threshold (and thus critical slowing down will no longer be optimally suppressed), one still has the advantage that not every individual spin has to be considered for inclusion in the cluster. Finally, it would be interesting to see whether this algorithm can be useful in other fields of physics where cluster algorithms have come to flourish.

Acknowledgements

I wish to thank Henk Blöte and Kurt Binder for the fruitful collaboration that has led to the work presented in this paper. I am grateful to Holger Meßingfeld for producing the data shown in Fig. 1 and performing the least-squares fits for the inverse-square chain.

References

1. E. Luijten and H. W. J. Blöte, Int. J. Mod. Phys. C **6**, 359 (1995).
2. R. H. Swendsen and J.-S. Wang, Phys. Rev. Lett. **58**, 86 (1987).
3. U. Wolff, Phys. Rev. Lett. **62**, 361 (1989).
4. P. W. Kasteleyn and C. M. Fortuin, J. Phys. Soc. Jpn. Suppl. **26s**, 11 (1969).
5. C. M. Fortuin and P. W. Kasteleyn, Physica **57**, 536 (1972).
6. E. Luijten and H. W. J. Blöte, Phys. Rev. B **56**, 8945 (1997).
7. E. Luijten and H. Meßingfeld, in preparation.
8. J. F. Nagle and J. C. Bonner, J. Phys. C **3**, 352 (1970).
9. P. W. Anderson and G. Yuval, J. Phys. C **4**, 607 (1971).
10. B. G. S. Doman, Phys. Stat. Sol. (b) **103**, K169 (1981).
11. G. V. Matvienko, Theor. Math. Phys. **63**, 635 (1985).
12. J. O. Vigfusson, Phys. Rev. B **34**, 3466 (1986).
13. M. Aizenman, J. T. Chayes, L. Chayes, and C. M. Newman, J. Stat. Phys. **50**, 1 (1988).
14. J. L. Monroe, R. Lucente, and J. P. Hourlland, J. Phys. A **23**, 2555 (1990).
15. M. J. Wragg and G. A. Gehring, J. Phys. A **23**, 2157 (1990).
16. R. Manieri, Phys. Rev. A **45**, 3580 (1992).
17. Z. Glumac and K. Uzelac, J. Phys. A **26**, 5267 (1993).
18. J. L. Monroe, J. Stat. Phys. **76**, 1505 (1994).
19. S. A. Cannas and A. C. N. de Magalhães, J. Phys. A **30**, 3345 (1997).
20. E. Luijten, Phys. Rev. E, to be published (see also cond-mat/9811332).
21. E. Luijten, H. W. J. Blöte, and K. Binder, Phys. Rev. Lett. **79**, 561 (1997).
22. E. Luijten, H. W. J. Blöte, and K. Binder, Phys. Rev. E **56**, 6540 (1997).
23. M. A. Anisimov, A. A. Povodyrev, V. D. Kulikov, and J. V. Sengers, Phys. Rev. Lett. **75**, 3146 (1995).
24. E. Luijten and K. Binder, in preparation.
25. Z. Glumac and K. Uzelac, Phys. Rev. E **58**, 4372 (1998).

Phase Diagram of the Spin-3/2 Blume-Capel Model

J. A. Plascak[1] and D. P. Landau[2]

[1] Universidade Federal de Minas Gerais, Departamento de Física ICEx, C.P. 702, CEP 30161-123, Belo Horizonte - MG, Brazil
[2] Center for Simulational Physics, The University of Georgia, Athens GA 30602, USA

Abstract. Monte Carlo simulations have been used to study the two-dimensional spin-3/2 Blume-Capel model. Numerical results have been obtained from simulations on finite square lattices with $L \times L$ sites and periodic boundary conditions in all directions where $4 \leq L \leq 192$. Qualitative results for the phase diagram, as well as high statistics simulations for zero crystal field, have been obtained by computing the magnetization, susceptibility and fourth-order cumulant using single histograms technique near the critical temperature. An isolated multiphase critical point is found at the terminus of the first-order transition line, and the universality class along the second-order transition line is the same as that of the simple Ising model.

1 Introduction

The Blume-Capel model was originally proposed to account for first-order phase transitions in magnetic systems [1,2]. The Hamiltonian can be written as

$$H = -J \sum_{<ij>} S_i S_j + D \sum_i S_i^2 , \tag{1}$$

where $J > 0$ is the ferromagnetic exchange interaction between nearest neighbor spins and D is the single ion anisotropy. The first sum is taken over nearest-neighbors pairs of spins and the second sum is taken over the N spins on a d-dimensional lattice. The spin variables S_i, in the general case, assume values $-S, -S+1, ..., S-1, S$. In one dimension the above model has no phase transition at finite temperature for any value of S and D. The $S = 1$ case has been extensively studied by several approximate techniques in two- and three-dimensions and its phase diagram is well established [1–9]. For greater values of spin, however, fewer results are available in the literature: mean field approximation [10], Cayley tree formulation [11], finite size scaling based on transfer matrix solution [12] and Monte Carlo simulations [13] furnish a phase diagram where an isolated multiphase critical point is immersed in the ferromagnetic ordered phase. However, other Monte Carlo simulations [14], renormalization group [15] and finite size renormalization group calculations [9] predict the appearance of a tetracritical point instead.

Springer Proceedings in Physics, Volume 85
Computer Simulation Studies in Condensed-Matter Physics XII
Eds.: D.P. Landau, S.P. Lewis and H.-B. Schüttler

In this work we present new Monte Carlo simulations for the $d = 2$ case and examine the behavior of the $S = 3/2$ model.

2 Monte Carlo Simulations

Our simulations have been done on finite $L \times L$ lattices with periodic boundary conditions in both directions. We have used the standard Metropolis single spin flip algorithm and the lattice sites have been visited in a sequential way. Besides the energy, we have computed the magnetization

$$M = \frac{1}{L^2} \sum_i S_i \,, \tag{2}$$

and its derivatives such as the zero field magnetic susceptibility χ as well as the fourth-order cumulant U, and the quadrupole moment

$$Q = \frac{1}{L^2} \sum_i S_i^2 \,, \tag{3}$$

together with its quadrupolar susceptibility χ_Q. Three different numerical analyses have been performed.

2.1 Test Simulation

A simulation has been done for $L = 4$ where the numerical results are compared to the exact ones obtained by enumeration of all states. We have used a total of 2.42×10^6 Monte Carlo steps per spin and discarded the first 2×10^4 steps for thermalization. Single histograms have been constructed for a given reduced temperature $k_B T/J$ where k_B is the Boltzmman constant. In Fig. 1 we plot the exact results together with the simulations at $D = 0$ for the energy E, the fourth-order cumulant U, the mean value of the modulus of the magnetization $|M|$ and the mean value for Q. The results were taken from histograms at $k_B T/J = 3.0$, however there is no noticeable difference if we take histograms at greater temperatures (for instance, $k_B T/J = 5.0$).

2.2 High Statistics Simulations

High statistics simulations have been performed at $D = 0$ for lattices varying from $L = 12$ to $L = 192$. Histograms at temperature $k_B/J = 3.291$ have been obtained by running 3.02×10^6 Monte Carlo steps per spin and discarding the first 2×10^4 steps. The results for the fourth-order cumulant are shown in Fig. 2. The critical temperature is located at $k_B/J = 3.2900(5)$ which is comparable to the estimate $k_B T/J = 3.287(2)$ obtained from the transfer-matrix finite size scaling [12]. At this temperature we estimate $U^* = 0.61(1)$ which is consistent with the value expected for the Ising universality class.

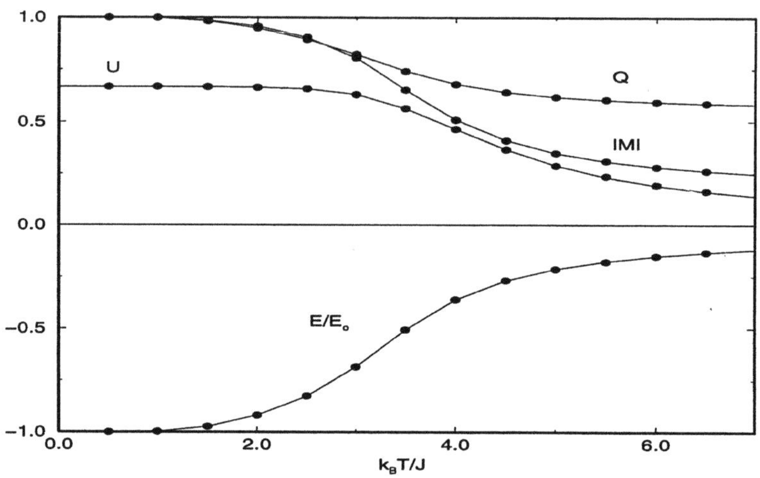

Fig. 1. Monte Carlo simulations (dots) and exact results (full lines) for $L = 4$ at $D = 0$. The statistic errors are much smaller than the size of the dots

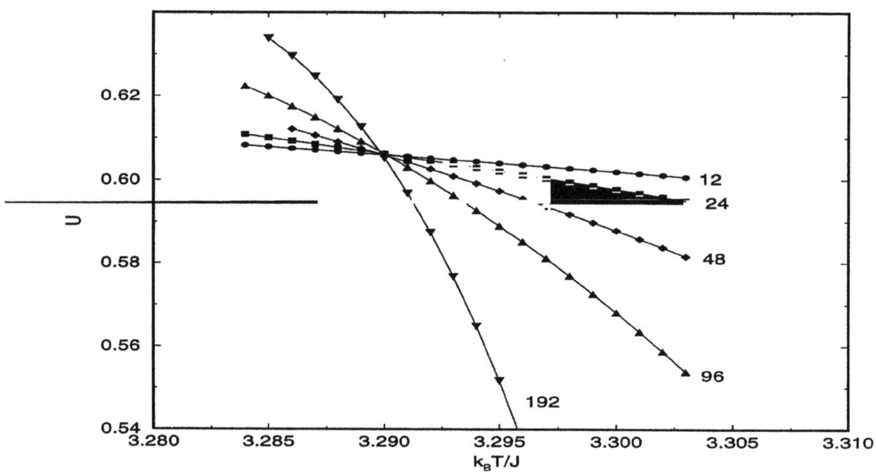

Fig. 2. Fourth-order cumulant as function of the temperature for different lattices sizes. The lines are guide to the eyes. The error in the cumulant varies from 0.0008 for the smallest lattice to order of 0.01 for the larger ones at the temperature where the histograms have been taken

2.3 Phase Diagram

We have performed more qualitative simulations in order to get a general view of the phase diagram near the region of the first-order transition line. For lattice sizes $L = 64$ and $L = 128$ we have computed the thermodynamic quantities by taking a total of 6×10^4 Monte Carlo steps per spin and discarding the first 10^4 for thermalization. The first-order transition lines have been located through the discontinuity in the magnetization while the second-order one through the maximum value of the magnetic susceptibility χ. The results are shown in Fig. 3. It is clear from this figure that, instead of a tetracritical point, an isolated critical point appears at the terminus of the first-order transition line. Simulations in the region closer to $D/4J = 0.5$ indicate that the second-order transition line does not present, as it may appear in the figure, any kink (the same applies for the first-order line in the same region). Exact location of phase boundaries requires a finite size scaling analysis. This is currently being done.

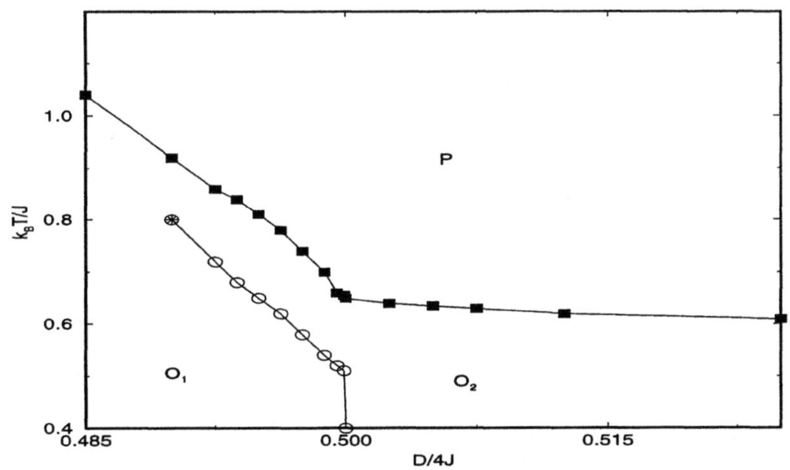

Fig. 3. Phase diagram for simulations done on a finite lattice with $L = 64$ in the region close to the first-order transition. Similar results are obtained for $L = 128$. O_1 and O_2 represent ordered phases with different magnetizations and P represents the paramagnetic phase. The squares are second-order phase transitions while circles are first-order phase transitions. The lines are guide to the eyes. The error in locating the transitions are roughly the size of the symbols

3 Summary

We have studied the two-dimensional $S = 3/2$ Blume-Capel model by Monte Carlo simulations. Our preliminary results show that the first-order transition line ends up in an isolated multiphase critical point instead of a tetracritical point. As expected, the second-order transition line separating the ordered phases from the disordered one is in the same universality class as the two-dimensional Ising model.

Acknowledgements

JAP would like to thank the financial support from CAPES, CNPq, FAPEMIG and FINEP (Brazilian agencies) and PRONEX for financial support. He would also like to thank the Center for Simulational Physics (UGA, Athens - Georgia) for the kind hospitality when part of this work has been done. The research was also supported in part by NSF grant DMR-9727714.

References

1. M. Blume, Phys. Rev. **141** 517 (1966).
2. H. W. Capel, Physica **32** 966 (1966).
3. M. Blume, V. J. Emery and R. B. Griffiths, Phys. Rev. **A4** 1071 (1971).
4. D. M. Saul, M. Wortis and D. Staufer, Phys. Rev. B**9** 4964 (1974).
5. A. K. Jain and D. P. Landau, Phys. Rev. B**22** 445 (1980).
6. O. F. de Alcantara Bonfim and C. H. Obcemea, Z. Phys. **B64** 469 (1986).
7. A. N. Berker and M. Wortis, Phys. Rev. **B14** 4946 (1976).
8. O. F. de Alcantara Bonfim, Physica **A130** 367 (1985).
9. S. Moss de Oliveira, P. M. C. de Oliveira and F. C. Sá Barreto, J. Stat. Phys. **78** 1619 (1995).
10. J. A. Plascak, J. G. Moreira and F. C. Sá Barreto, Phys. Lett. A **173** 360 (1993).
11. M. N. Tamashiro and S. R. Salinas, Physica **A211** 124 (1994).
12. J. C. Xavier, F. C. Alcaraz, D. Peña Lara and J. A. Plascak, Phys. Rev. B **57** 11575 (1998).
13. D. Peña Lara and J. A. Plascak, Int. J. Mod. Phys. B **12** 2045 (1998).
14. F. C. Sá Barreto and O. F. Alcantara Bonfim, Physica**A172** 378 (1991).
15. A. Bakchich, A. Bassir and A. Benyoussef, Physica **A195** 188 (1993).

Dynamic Phase Transition and Hysteresis in Kinetic Ising Models

P. A. Rikvold[1,2], G. Korniss[2], C. J. White[1,2], M. A. Novotny[2], and S. W. Sides[3]

[1] Department of Physics and Center for Materials Research and Technology, Florida State University, Tallahassee, FL 32306-4350, USA
[2] Supercomputer Computations Research Institute, Florida State University, Tallahassee, FL 32306-4130, USA
[3] Integrated Materials Research Laboratory, Sandia National Laboratory, Albuquerque, NM 87123, USA

Abstract. We briefly introduce hysteresis in spatially extended systems and the dynamic phase transition observed as the frequency of the oscillating field increases beyond a critical value. Hysteresis and the decay of metastable phases are closely related phenomena, and a dynamic phase transition can occur only for field amplitudes, temperatures, and system sizes at which the metastable phase decays through nucleation and growth of *many* droplets. We present preliminary results from extensive Monte Carlo simulations of a two-dimensional kinetic Ising model in a square-wave oscillating field and estimate critical exponents by finite-size scaling techniques adapted from equilibrium critical phenomena. The estimates are consistent with the universality class of the two-dimensional equilibrium Ising model and inconsistent with two-dimensional random percolation. However, we are not aware of any theoretical arguments indicating why this should be so. Thus, the question of the universality class of this nonequilibrium critical phenomenon remains open.

1 Introduction

Hysteresis often occurs when a bistable or multistable system is driven by an oscillating force which varies too fast for the system to respond without a phase lag. In fact, the word was coined by Ewing from the Greek *husterein* ($\overset{c}{\upsilon}\sigma\tau\epsilon\rho\acute{\epsilon}\omega$) which means "to be behind" [1]. Examples include ferromagnets and ferroelectrics in AC fields, electrochemical cyclic-voltammetry experiments, and nonlinear elastic media under oscillating stress, just to mention a few. The most familiar representation of hysteresis is probably the (usually) closed curve obtained by plotting the system response versus the applied force. Examples of such *hysteresis loops* are shown in Fig. 1(a). This figure shows data from a Monte Carlo (MC) simulation of a two-dimensional kinetic Ising ferromagnet below its critical temperature, which is driven by a sinusoidally oscillating field. However, the shape of the loops shown is quite general. For simplicity and concreteness, in this paper we use magnetic language, designating the oscillating force the "field" and the system response

Springer Proceedings in Physics, Volume 85
Computer Simulation Studies in Condensed-Matter Physics XII
Eds.: D.P. Landau, S.P. Lewis and H.-B. Schüttler

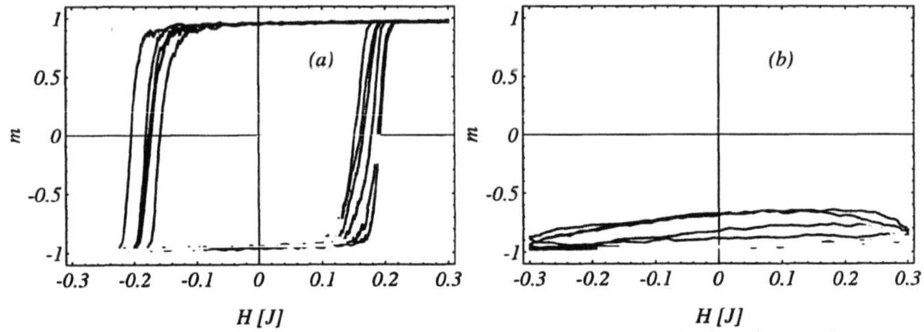

Fig. 1. Series of hysteresis loops obtained from a Monte Carlo (MC) simulation of a two-dimensional kinetic Ising ferromagnet at $T = 0.8\ T_c$ in a sinusoidally varying field. The dimensionless magnetization m is shown vs. the field H in units of the nearest-neighbor interaction energy J. (a): A series of regular symmetric hysteresis loops, typical of a system driven at a moderately low frequency. (b): A series of asymmetric hysteresis loops, typical of a system driven at a high frequency

the "magnetization". These quantities can easily be re-interpreted when one discusses other types of systems, such as those mentioned above.

Hysteresis was first systematically investigated in the late 19th century by engineers and physicists primarily concerned with the development of electric motors and transformers [1–5]. The area of the hysteresis loop is proportional to the magnetic energy loss during one field period, as was first pointed out by Warburg [2]. Its dependences on the frequency and amplitude of the applied field have therefore been studied intensively ever since, but open questions still remain. In particular, the question of whether the low-frequency behavior for ultrathin films of highly anisotropic materials is asymptotically a power law or logarithmic is still under investigation, both experimentally [6–10] and theoretically [11–17].

A different aspect of hysteresis in bistable systems, which is the main topic of the present paper, occurs at higher driving frequencies. When the frequency becomes sufficiently high, the symmetry of the hysteresis loop, which is apparent in Fig. 1(a), is broken. Instead of oscillating between its two stable values with a phase lag relative to the field, the magnetization oscillates around one or the other of its zero-field stable values. A series of such asymmetric loops is shown in Fig. 1(b). This symmetry breaking has become a topic of vigorous research during the last two decades. It was first reported by Tomé and de Oliveira [18], who observed it during numerical solutions of a mean-field equation of motion for a ferromagnet in an oscillating field.

Subsequently, symmetry breaking has been observed in numerous MC simulations of kinetic Ising systems, [17,19–24], as well as in further mean-field studies [20,22,23,25]. It may also have been experimentally observed in ultrathin films of Co on Cu(001) [7,8]. Reviews of the field as it stood in 1994 and as it stands today can be found in [26] and [27], respectively. There now appears to be a consensus that the symmetry breaking corresponds to a genuine second-order, nonequilibrium phase transition. Associated with the transition is a divergent time scale (critical slowing-down) [22] and, for spatially extended systems, a divergent correlation length [17,24]. Although estimates of the critical exponents have recently become available from finite-size scaling analyses of MC data for a two-dimensional Ising system in a sinusoidally oscillating field [17,24], their accuracy is not yet sufficient to decide whether the dynamic phase transition in this system belongs to a previously known universality class or represents a new one. Neither are we aware of theoretical arguments that can resolve the issue or help to identify the correct "critical clusters".

The main purpose of this paper is to present preliminary results for two-dimensional kinetic Ising systems in square-wave oscillating fields. These may help point the way towards answering some of the remaining questions concerning this intriguing nonequilibrium critical phenomenon. The use of a square-wave field both tests the universality of the dynamic phase transition and significantly increases the computational speed. The structure of the remainder of the paper is as follows. The kinetic model system and the relevant quantities, including the dynamic order parameter, are defined in Sect. 2. A brief primer on metastable decay in spatially extended systems is given in Sect. 3. Numerical results are presented in Sect. 4, and a discussion and some suggestions for further research are given in Sect. 5.

2 Model and Relevant Quantities

Most numerical simulations of the dynamic phase transition (hereafter abbreviated DPT) in spatially extended systems have been performed on nearest-neighbor kinetic Ising ferromagnets on hypercubic lattices with periodic boundary conditions. These models are defined by the Hamiltonian

$$\mathcal{H} = -J \sum_{\langle ij \rangle} s_i s_j - H(t) \sum_i s_i \ , \tag{1}$$

where $s_i = \pm 1$ is the state of the ith spin, $\sum_{\langle ij \rangle}$ runs over all nearest-neighbor pairs, $J > 0$ is the ferromagnetic interaction, $\sum_i$ runs over all L^d lattice sites, and $H(t)$ is an oscillating, spatially uniform applied field. The magnetization per site is

$$m(t) = L^{-d} \sum_{i=1}^{L^d} s_i(t) \ . \tag{2}$$

The temperature T is fixed below its zero-field critical value T_c, so that the magnetization for $H = 0$ has two degenerate spontaneous equilibrium values, $\pm m_{sp}(T)$. For nonzero fields the equilibrium magnetization has the same sign as H, while for not too strong H (see Sect. 3 for quantitative statements) the opposite magnetization direction becomes *metastable* and decays slowly towards equilibrium with time.

The dynamic used here, as well as in [17] and [24], is the Glauber single-spin-flip MC algorithm with updates at randomly chosen sites [28]. The time unit is one MC step per spin (MCSS). Each attempted spin flip from s_i to $-s_i$ is accepted with probability

$$W(s_i \to -s_i) = \frac{\exp(-\beta \Delta E_i)}{1 + \exp(-\beta \Delta E_i)} \, . \tag{3}$$

Here ΔE_i is the energy change that would result from accepting the spin flip, and $\beta = 1/k_{\mathrm{B}}T$ where k_{B} is Boltzmann's constant. For the largest system ($L=512$ square lattice) we employed a massively parallel implementation of this algorithm [29–32]. Other dynamics that have been used in MC studies of the DPT are Glauber or Metropolis [28] with updates at sequentially selected sites [20–23,26,27]. Although the choice of update scheme can lead to subtle differences in the dynamics [33] and we prefer random site selection as the more physical scheme, we do not believe it affects universal aspects of the DPT.

The dynamic order parameter is the period-averaged magnetization [18],

$$Q = \frac{1}{2t_{1/2}} \oint m(t) \; \mathrm{d}t \, , \tag{4}$$

where $t_{1/2}$ is the half-period of the oscillating field, and the beginning of the period is chosen at a time when $H(t)$ changes sign. Analogously we also define the local order parameter

$$Q_i = \frac{1}{2t_{1/2}} \oint s_i(t) \; \mathrm{d}t \, , \tag{5}$$

which is the period-averaged spin at site i. For slowly varying fields the probability distribution of Q is sharply peaked at zero [17]. We shall refer to this as the *dynamically disordered phase*. For rapidly oscillating fields the Q distribution becomes bimodal with two sharp peaks near $\pm m_{\mathrm{sp}}(T)$, corresponding to the broken symmetry of the hysteresis loops [17]. We shall refer to this as the *dynamically ordered phase*. Near the DPT we use finite-size scaling analysis of MC data to estimate the critical exponents that characterize the transition.

Previous studies of the DPT have used an applied field which varies sinusoidally in time. While sinusoidal or linear saw-tooth fields are the most common in experiments and are necessary to obtain a vanishing loop area in the low-frequency limit [16,17], the wave form of the field should not affect universal aspects of the DPT. This should be so because the transition essentially depends on the competition between two time scales: the half-period $t_{1/2}$ of the applied field, and the average time it takes the system to

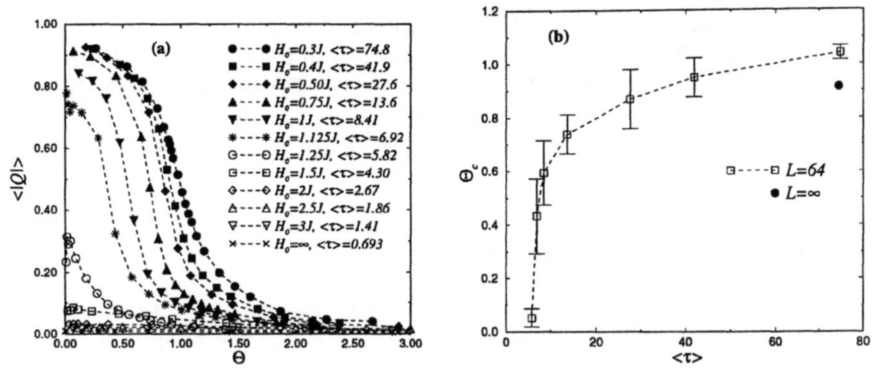

Fig. 2. **(a)** Order parameter $\langle|Q|\rangle$ vs. Θ for $L=64$ square system at $T=0.8T_c$ for several values of the field amplitude H_0. The corresponding metastable lifetimes, $\langle\tau(H_0,T)\rangle$ given in MCSS, are shown in the legends. **(b)** The approximate dimensionless critical half-period Θ_c vs. the metastable lifetime $\langle\tau(H_0,T)\rangle$ in MCSS

leave the metastable region near one of its two degenerate zero-field equilibria when a field of magnitude H_0 and sign opposite to the magnetization is applied. This *metastable lifetime*, $\langle\tau(H_0,T)\rangle$, is usually estimated as the average first-passage time to zero magnetization. In the present paper we use a *square-wave* field of amplitude H_0. This has significant computational advantages over the sinusoidal field variation since we can use two look-up tables to determine the acceptance probabilities: one for $H=+H_0$ and one for $H=-H_0$.

In terms of the dimensionless half-period,

$$\Theta = t_{1/2}/\langle\tau(H_0,T)\rangle \ , \qquad (6)$$

the DPT should occur at a critical value Θ_c of order unity. Although Θ can be changed by varying either $t_{1/2}$, H_0, or T, in a first approximation we expect Θ_c to be independent of H_0 and T. This expectation is confirmed by simulations carried out at several H_0 and T for different system sizes. In particular, Fig. 2(a) shows the average norm of the order parameter $\langle|Q|\rangle$ vs. Θ for various field amplitudes and the corresponding metastable lifetimes. For weaker fields (longer lifetimes) the transition is apparent at some $\Theta_c \sim 1$, while it clearly disappears (no dynamically ordered phase exists) for sufficiently strong fields (small lifetimes). Figure 2(b) shows Θ_c vs. $\langle\tau(H_0,T)\rangle$ where Θ_c was determined approximately as the location of the peak in the fluctuations of Q in a $L=64$ system. Note that Θ_c for an infinite system is slightly different from this estimate due to finite-size effects. Also, these effects are expected to be smaller for strong fields (small lifetimes) where the spins become uncorrelated. The deviations from unity for very small $\langle\tau(H_0,T)\rangle$ and the complete disappearance of the DPT are discussed further in Sect. 3.

In many studies of the DPT the transition has been approached by changing H_0 or T [19–21,23]. While the above discussion indicates that this is correct in principle, we show in Sect. 3 that $\langle \tau(H_0, T) \rangle$ depends strongly and nonlinearly on its arguments. We therefore prefer changing $t_{1/2}$ at constant H_0 and T [17,24], as this in practice gives more precise control over the distance from the transition.

3 Decay of Metastable Phases

In mean-field studies of the DPT H_0 must be larger than the temperature-dependent spinodal field beyond which the metastable free-energy minimum disappears [11]. For weaker fields the hysteresis loops remain asymmetric, even for the lowest frequencies. It is much less appreciated that bounds on the fields and temperatures for which a DPT can occur also exist for spatially extended systems. These bounds are readily obtained from classical nucleation theory [34] and the Kolmogorov-Johnson-Mehl-Avrami (KJMA) theory of metastable decay [17,24,33,35–37] by comparing four characteristic lengths. These are the lattice constant (here defined as unity), the size $R_c(|H|, T)$ of a randomly nucleated critical droplet of stable phase, the typical distance $R_0(|H|, T)$ between individual droplets, and the system size L.

The critical radius is determined by the competition between the surface free energy of the droplet, $\propto \sigma(T) R^{d-1}$ where $\sigma(T)$ is the surface tension between the two phases, and the bulk free energy, $\propto -|H| R^d$. As a result, $R_c \propto \sigma(T)/|H|$. The nucleation rate per unit volume is $I(H, T) \propto \exp\left[-\Xi(T)/|H|^{d-1}\right]$, where $\Xi(T)$ is the field-independent part of the free energy of a critical droplet divided by $k_B T$, and preexponential powers of $|H|$ have been suppressed.

The classical KJMA theory describes the metastable decay as homogeneous nucleation of droplets at random times and positions with nucleation rate I, followed by deterministic growth of these droplets with constant radial velocity $v \propto |H|$. Assuming that the droplets overlap freely when they meet, the time dependent magnetization is given by the well-known "Avrami's Law",

$$
\begin{aligned}
m(t) &\approx m_{sp} \left\{ 2 \exp\left[-\int_0^t I \Omega_d (vt')^d \mathrm{d}t' \right] - 1 \right\} \\
&= m_{sp} \left\{ 2 \exp\left[-\frac{\Omega_d v^d I}{d+1} t^{d+1} \right] - 1 \right\} ,
\end{aligned}
\tag{7}
$$

where Ω_d is a proportionality constant such that the volume of a droplet of radius R equals $\Omega_d R^d$. The argument of the exponential is the "extended volume" [36], i.e., the total volume fraction of stable-phase droplets, *uncorrected* for overlaps. Solving (7) for the time at which $m = 0$ gives the lifetime,

$$
\langle \tau(|H|, T) \rangle = \left[\frac{\Omega_d v^d I}{(d+1) \ln 2} \right]^{-1/(d+1)} ,
\tag{8}
$$

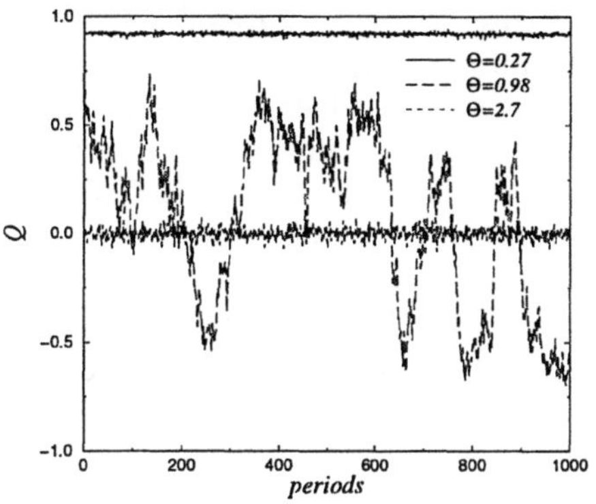

Fig. 3. Time series of the order parameter Q at $T=0.8T_c$ and $H_0=0.3J$ for $L = 128$. Horizontal trace near $Q = +1$: $\Theta = 0.27 < \Theta_c$ (dynamically ordered phase). Strongly fluctuating trace: $\Theta = 0.98 \approx \Theta_c$ (near the DPT). Horizontal trace near $Q = 0$: $\Theta = 2.7 > \Theta_c$ (dynamically disordered phase)

which depends on $|H|$ and T through v and I, but is *independent* of L. The characteristic length R_0 is obtained from $v(|H|,T)$ and $\langle\tau(|H|,T)\rangle$ as

$$R_0(|H|,T) = v(|H|,T)\ \langle\tau(|H|,T)\rangle \propto \exp\left[\frac{\Xi(T)}{(d+1)|H|^{d+1}}\right], \tag{9}$$

where a preexponential power of $|H|$ again has been suppressed [33,35].

The regime in which a DPT can occur is that in which a large number of droplets contribute to the decay of the metastable phase:

$$1 \ll R_c \ll R_0 \ll L\,. \tag{10}$$

This is known as the multidroplet (MD) regime [33,35]. It is limited on the weak-field/small-system side by the *dynamic spinodal* (DSP) field ,

$$H_{\mathrm{DSP}}(T,L) \sim \left(\frac{1}{d+1}\frac{\Xi(T)}{\ln L}\right)^{1/(d-1)}, \tag{11}$$

which corresponds to $R_0 \approx L$. For $|H| < H_{\mathrm{DSP}}$ almost always only a *single* droplet contributes to the magnetization switching. A DPT does *not* occur in this regime, but *stochastic resonance* is observed at low frequencies [38]. On the strong-field side the MD regime is limited by an L-independent crossover field approximately given by $2R_c \approx 1$. By a somewhat confusing term this crossover is often referred to as the "mean-field spinodal" $H_{\mathrm{MFSP}}(T)$ [33,35]. In the *strong-field* (SF) regime beyond this limit the spins become increasingly uncorrelated as $|H|$ increases. This is the region of small $\langle\tau(H_0,T)\rangle$ (large H_0) where Θ_c rapidly approaches zero and the transition disappears (Fig. 2). In computational studies of the DPT it is essential to ensure that $H_{\mathrm{DSP}}(T,L) < H_0 < H_{\mathrm{MFSP}}(T)$ for all values of L and T used.

4 Results

We performed extensive simulations on square lattices with L between 64 and 512 at $T=0.8T_c$ and $H_0=0.3J$. Typical run lengths near the transition range from 2×10^4 to 10^5 full periods, corresponding to 2.8×10^6 – 1.4×10^7 MCSS. The system was initialized with all spins up and and the square-wave external field started with the half-period in which $H = -H_0$. After some relaxation the system magnetization reaches a limit cycle (except for thermal fluctuations), i.e., Q is stationary. We discarded the first 500 periods of the time series to exclude transients from the stationary-state averages.

For small half-periods ($\Theta \ll \Theta_c$) the magnetization does not have time to switch, resulting in $|Q| \approx m_{sp}$, while for large half-periods ($\Theta \gg \Theta_c$) it switches every half-period and $Q \approx 0$ as can be seen from the time series in Fig. 3. The transition between the high- and low-frequency regimes is singular, characterized by large fluctuations in Q. To illustrate the spatial aspects of the transition we also show configurations of the local order parameter Q_i in Fig. 4. Below Θ_c [Fig. 4(a)] the majority of spins spend most of their time in the +1 state, i.e., in the metastable phase during the first half-period, and in the stable equilibrium phase during the second half-period (except for equilibrium fluctuations). Thus, most of the $Q_i \approx +1$. Droplets of $s_i = -1$ that nucleate during the negative half-period and then decay back to +1 during the positive half-period show up as roughly circular gray spots in the figure. Since the spins near the center of such a droplet become negative first and revert to positive last, these spots appear darkest in the middle. Above Θ_c [Figs. 4(c,d)] the system follows the field in every half-period (with some phase lag) and $Q_i \approx 0$ at all sites i. Near Θ_c [Fig. 4(b)] there are large clusters of both $Q_i \approx +1$ and $Q_i \approx -1$ separated by wide "interfaces" where $Q_i \approx 0$. Also, for not too large lattices one often observes the full reversal of an ordered configuration $\{Q_i\} \to -\{Q_i\}$, typical of finite, spatially extended systems undergoing symmetry breaking.

For finite systems in the dynamically ordered phase the distribution of Q becomes bimodal. Thus, to capture symmetry breaking, one has to measure the average norm of Q as the order parameter, i.e., $\langle|Q|\rangle$. Figure 5(a) clearly shows that this order parameter is of order unity for $\Theta < \Theta_c$ and vanishes for $\Theta > \Theta_c$, except for finite-size effects.

To characterize and quantify this transition in terms of critical exponents we employ the well-known technique of finite-size scaling [28,39]. The quantity analogous to the susceptibility is the scaled variance of the dynamic order parameter,

$$X_L = L^2 \left(\langle Q^2\rangle_L - \langle|Q|\rangle_L^2\right) \ . \tag{12}$$

For finite systems X_L has a characteristic peak near Θ_c [see Fig. 5(b)] which increases in height with increasing L, while no finite-size effects can be observed for $\Theta \ll \Theta_c$ and $\Theta \gg \Theta_c$. This implies the existence of a divergent length scale, possibly the correlation length which governs the long-distance behavior of the local order-parameter correlations $\langle Q_i Q_j\rangle$. Note that the location of the maximum in X_L shifts with L. This also contains important information about the critical exponents.

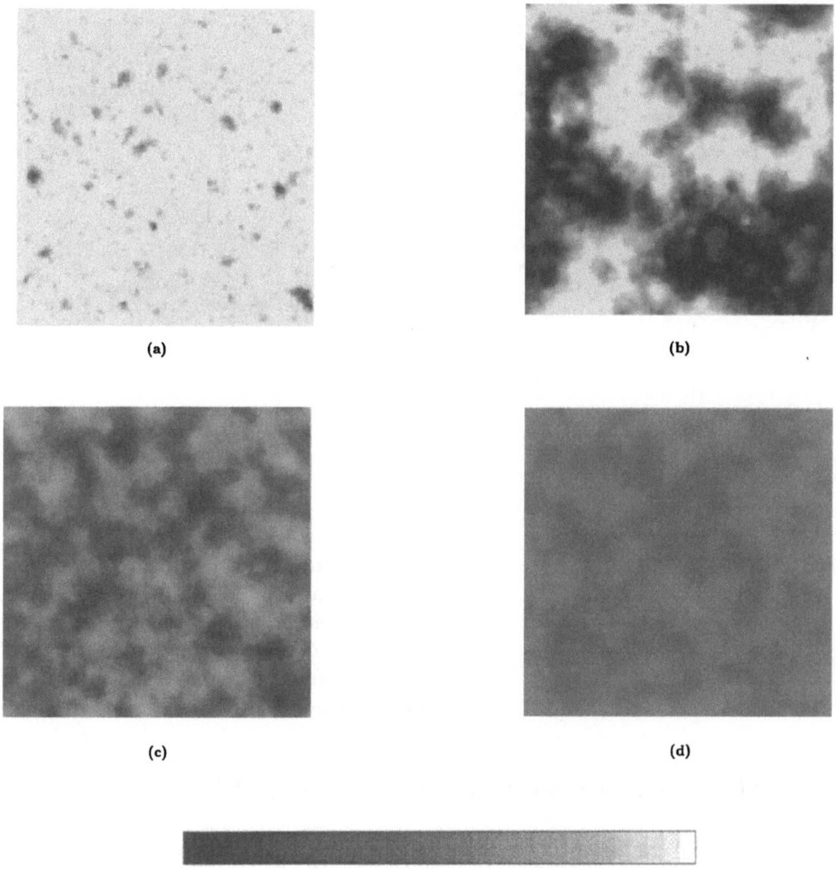

Fig. 4. Configurations of the local order parameter Q_i at $T = 0.8T_c$ and $H_0 = 0.3J$ for $L = 128$. **(a)** $\Theta = 0.27 < \Theta_c$ (dynamically ordered phase). **(b)** $\Theta = 0.98 \approx \Theta_c$ (near the DPT). **(c)** $\Theta = 2.7 > \Theta_c$ (dynamically disordered phase). **(d)** $\Theta = 13.4 \gg \Theta_c$ (deeper in the disordered phase). On the gray-scale -1 ($+1$) corresponds to black (white)

To estimate the value of Θ_c at which the transition occurs in an infinite system we use the intersection of the fourth-order cumulant ratios [28,39],

$$U_L = 1 - \frac{\langle Q^4 \rangle_L}{3\langle Q^2 \rangle_L^2} , \tag{13}$$

for several system sizes as shown in Fig. 5(c). For the largest system ($L = 512$) the error bars on U_L were too large to use it to obtain estimates for the crossing. Our estimate for the dimensionless critical half-period is $\Theta_c =$

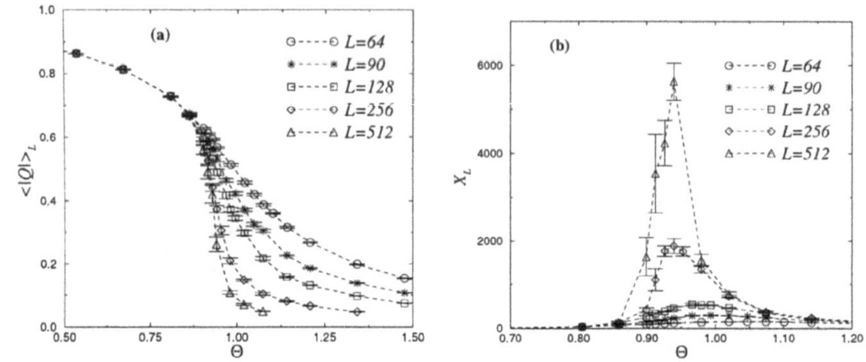

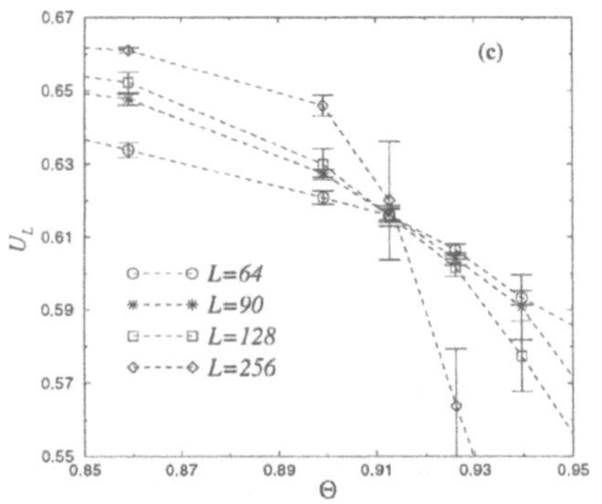

Fig. 5. Finite-size behavior of our observables at $T = 0.8T_c$ and $H_0 = 0.3J$ for various system sizes. **(a)** The order parameter $\langle|Q|\rangle_L$. **(b)** The scaled variance X_L as defined in (12). **(c)** The fourth-order cumulant ratio as defined in (13)

0.913 ± 0.003 with a fixed-point value $U^* = 0.615 \pm 0.005$ for the cumulant ratio.

For our model the quantity analogous to the reduced temperature in equilibrium systems (i.e., the distance from the critical point) is

$$\theta = \frac{|\Theta - \Theta_c|}{\Theta_c} . \tag{14}$$

Finite-size scaling theory provides simple relations for the order parameter and its scaled variance X_L for finite systems in the critical regime [28,39]:

$$\langle|Q|\rangle_L = L^{-\beta/\nu}\mathcal{F}_\pm(\theta L^{1/\nu}) , \tag{15a}$$
$$X_L = L^{\gamma/\nu}\mathcal{G}_\pm(\theta L^{1/\nu}) , \tag{15b}$$

where $\mathcal{F}_\pm$ and $\mathcal{G}_\pm$ are scaling functions and the $+$ ($-$) index refers to $\Theta > \Theta_c$ ($\Theta < \Theta_c$). Then at Θ_c ($\theta = 0$) we have the finite-size behavior of the above two observables:

$$\langle|Q|\rangle_L \propto L^{-\beta/\nu} , \tag{16a}$$
$$X_L \propto L^{\gamma/\nu} . \tag{16b}$$

Using $\langle|Q|\rangle_L$ and X_L for several system sizes at Θ_c (estimated as the value of Θ where the cumulants cross) we employ (16a) and (16b) to find the exponent ratios β/ν and γ/ν through weighted linear least-squares fitting to the logarithmic data. Further, the shift in the location of the peak in X_L for these finite systems, $\Theta_c(L)$, is used to estimate ν in the same way [28,39]:

$$|\Theta_c(L) - \Theta_c| \propto L^{-1/\nu} . \tag{17}$$

Our estimates for these exponents are β/ν=0.122 ± 0.005, γ/ν=1.77 ± 0.05, and ν=1.0 ± 0.15. The largest uncertainty occurs in ν due to the large relative errors in $|\Theta_c(L) - \Theta_c|$. We note that these numbers are very close (within the one-standard-deviation error bars) to the critical exponents of the two-dimensional Ising universality class β/ν=1/8=0.125, γ/ν=7/4=1.75, and ν=1. However, as discussed in Sect. 5, we do not consider this conclusive proof of the universality class for this nonequilibrium phase transition.

To obtain a more general picture of how well the scaling relations in (15a) and (15b) are obeyed, we plot $\langle|Q|\rangle_L L^{\beta/\nu}$ [Fig. 6(a)] and $X_L L^{-\gamma/\nu}$ [Fig. 6(b)] vs. $\theta L^{1/\nu}$ [40]. In these figures we used the exponents of the two-dimensional Ising universality class since they are within one standard deviation of our estimates and analogous data-collapse plots using our numerical exponent estimates do not look perceptibly different. The plots show excellent agreement with the scaling assumption and graphically define the scaling functions $\mathcal{F}_\pm$ and $\mathcal{G}_\pm$. The asymptotic behaviors of these functions have to be simple power laws so that the true critical behavior is restored when the limit $L \to \infty$ is taken in (15a) and (15b). Some deviation can be observed in Fig. 6(b): below Θ_c the data points for the smaller systems systematically start to peel off earlier from the straight line representing the asymptotic behavior of $\mathcal{G}_-$ for large argument.

In the absence of theoretical arguments why this DPT should belong to the Ising universality class we also tested the data collapse with the exponents for random percolation, which are relatively close to the corresponding Ising ones [41]: β/ν=5/48≈0.104, γ/ν=43/24≈1.79, and ν=4/3≈1.33. It appears reasonable to consider this universality class in particular, since the metastable decay process described in Sect. 3 produces transient spanning clusters that belong to the random-percolation class as $m(t)$ passes through H- and T-dependent percolation thresholds for the two phases near $m = 0$ [42]. Data-collapse plots using the random-percolation exponents are shown in Figs. 6(c,d). They are clearly inferior to the ones with the Ising exponents [Figs. 6(a,b)]. In particular, for $\Theta > \Theta_c$ (the dynamically disordered phase) the data collapse gets progressively worse for larger systems. Comparing the scaling plots in Figs. 6(a,b) with Figs. 6(c,d) we conclude that the exponent ν for the DPT is significantly different from that of the random percolation universality class and that it is closer to the Ising value. This conclusion is supported by direct comparison of the random-percolation and Ising exponents with our numerical estimates for the DPT. Our estimates for β/ν and ν lie more than two standard deviations away from the percolation values,

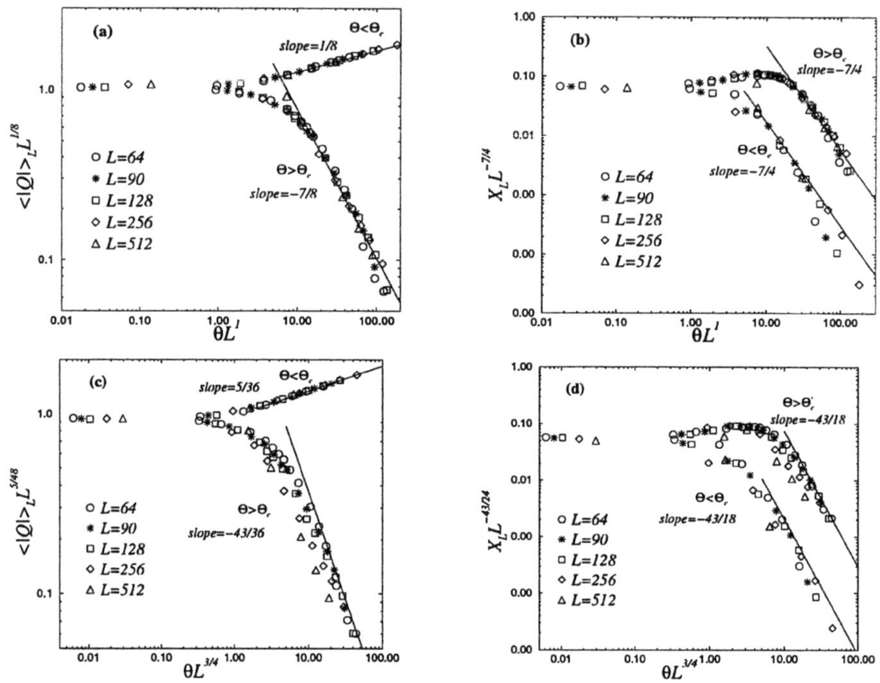

Fig. 6. Finite-size scaling plots at $T = 0.8T_c$ and $H_0 = 0.3J$ using the two-dimensional Ising exponents: **(a)** For the order parameter $\langle|Q|\rangle_L$. **(b)** For the scaled variance X_L. Finite-size scaling plots at the same temperature and field using the random-percolation exponents: **(c)** For the order parameter $\langle|Q|\rangle_L$. **(d)** For the scaled variance X_L. The straight lines in all graphs represent the asymptotic large-argument behaviors of the functions $\mathcal{F}_\pm$ and $\mathcal{G}_\pm$ given in (15a) and (15b), respectively. The value for the (infinite-system) Θ_c, used in all four graphs, was obtained from the cumulant crossing [(13)] Fig. 5(c)

but less than one standard deviation away from the Ising values, while our estimate for γ/ν lies midway between and less than one standard deviation away from the exact values for both classes.

The exponents previously obtained with a sinusoidally oscillating field [17,38] lie within two standard deviations of our values. As it was difficult to determine the uncertainties in the estimates obtained in that study, we believe the results are consistent and indicate that the DPT in systems driven by sinusoidal and square-wave fields belong to the same universality class, as expected.

5 Summary and Discussion

In this paper we have presented preliminary results from a large-scale Monte Carlo study of the dynamic phase transition (DPT) in a two-dimensional kinetic Ising ferromagnet driven by a square-wave oscillating field. Our results are consistent with those of previous studies of the same model in a sinusoidally varying field [17,24]. They indicate that for field amplitudes H_0 such that the metastable magnetized phase decays to equilibrium via the multidroplet (MD) mechanism described in Sect. 3, the system undergoes a continuous DPT when the half-period of the field, $t_{1/2}$, approximately equals the metastable lifetime, $\langle\tau(H_0,T)\rangle$. Thus the critical value Θ_c of the dimensionless half-period defined in (6) is near unity. As Θ is increased past Θ_c the order parameter $\langle|Q|\rangle$, which is the expectation value of the norm of the period-averaged magnetization, vanishes in a singular fashion, as shown in Fig. 5(a).

The strong and systematic finite-size effects in the order parameter and its scaled variance indicate that there is a divergent correlation length associated with the transition. Using standard finite-size scaling techniques borrowed from the theory of equilibrium phase transitions we therefore estimated Θ_c and the critical exponents β, γ, and ν from data for system sizes between $L=$ 64 and 512. The resulting estimates are $\beta/\nu=0.122\pm0.005$, $\gamma/\nu=1.77\pm0.05$, and $\nu=1.0\pm0.15$, which agree to within the statistical errors with those previously obtained with a sinusoidally oscillating field [17,38]. This is strong evidence that the shape of the wave does not effect the universal aspects of the DPT.

We also note that our estimates agree well within one standard deviation with the exact values for the two-dimensional equilibrium Ising model: $\beta/\nu=1/8=0.125$, $\gamma/\nu=7/4=1.75$, and $\nu=1$, and that they satisfy the exponent relation

$$2(\beta/\nu)+(\gamma/\nu)=2.01\pm0.05\approx d\,. \tag{18}$$

The fixed-point value of the fourth-order cumulant ratio, $U^*=0.615\pm0.005$, is also close to that of the Ising model, $U^*=0.610\,690\,1(5)$ [43]. As discussed in [17] we find the estimated values for the exponents and U^* conclusive evidence that the DPT corresponds to a nontrivial fixed point, so that (18) indeed represents a hyperscaling relation.

Since we know of no convincing theoretical argument that this nonequilibrium phase transition should be in any particular equilibrium universality class, we also compare the exponent estimates with those of the universality class of random percolation. For this class the values [41] $\beta/\nu=5/48\approx0.104$ and $\nu=4/3\approx1.33$ are significantly different from our exponent estimates, while $\gamma/\nu=43/24\approx1.79$ lies within the uncertainty in our estimate. Taken together, the three estimates and the low quality of the data collapse plots shown in Figs. 6(c,d) constitute strong evidence that the DPT is *not* in the universality class of random percolation. However, we emphasize that we do not yet consider the question of the universality class as settled. That will require still more accurate numerical exponent estimates as well as theoretical arguments.

Outside the MD regime the phase transition disappears. For weaker fields or smaller systems the metastable phase decays stochastically through a single droplet, and no long-range correlations evolve [38]. In the other extreme, as H_0 is increased into the strong-field regime, Θ_c goes to zero and vanishes at a sharply defined field (lifetime), as shown in Fig. 2(b).

In conclusion, the dynamic phase transition observed in spatially extended kinetic Ising systems driven by oscillating fields is a fascinating nonequilibrium critical phenomenon with all the hallmarks of a thermal phase transition corresponding to a nontrivial fixed point. Although the present study confirms that the particular wave form of the oscillating field does not change the universality class, and our numerical estimates for the critical exponents and cumulant ratio are consistent with the universality class of the two-dimensional Ising model in equilibrium, many questions remain. These include the question of universality under changes in T and H_0, and the details of the disappearance of the transition as H_0 is increased. No procedure to identify critical correlated clusters from the connected clusters seen in configuration "snapshots" such as Fig. 4(b) is yet known, and theoretical ideas are largely missing. Thus, there is still much to be done!

Acknowledgments

This research was supported in part by Florida State University through the Center for Materials Research and Technology and the Supercomputer Computations Research Institute (partially funded by the US Department of Energy through Contract No. DE-FC05-85ER25000), and by the National Science Foundation through Grants No. DMR-9634873 and DMR-9871455. Computing resources at the National Energy Research Scientific Computing Center were made available by the US Department of Energy under Contract No. DE-AC03-76SF00098.

References

1. Ewing J. A. (1881) Proc. R. Soc. London **33**, 21
2. Warburg E. (1881) Annalen der Physik und Chemie (Neue Folge) **13**, 141
3. Ewing J. A. (1882) Proc. R. Soc. London **34**, 39
4. Steinmetz C. P. (1892) Trans. Am. Inst. Electr. Eng. **9**, 3
5. Kline R. R. (1992) Steinmetz: Engineer and Socialist. Johns Hopkins University Press, Baltimore
6. He Y.-L., Wang G.-C. (1993) Phys. Rev. Lett. **70**, 2336
7. Jiang Q., Yang H.-N., Wang G.-C. (1995) Phys. Rev. B **52**, 14911
8. Jiang Q., Yang H.-N., Wang G.-C. (1996) J. Appl. Phys. **79**, 5122

9. Suen J. H., Erskine J. L. (1997) Phys. Rev. Lett. **78**, 3567
10. Suen J. H., Lee M. H., Teeter G., Erskine J. L. (1999) Phys. Rev. B **59**, 4249
11. Jung P., Gray G., Roy R., Mandel P. (1990) Phys. Rev. Lett. **65**, 1873
12. Rao M., Krishnamurthy H. R., Pandit R. (1990) Phys. Rev. B **42**, 856
13. Thomas P. B., Dhar D. (1993) J. Phys. A **26**, 3973
14. Luse C. N., Zangwill A. (1994) Phys. Rev. E **50**, 224
15. Beale P. D. (1994) Integrated Ferroelectrics **4**, 107
16. Sides S. W., Rikvold P. A., Novotny M. A. (1998) J. Appl. Phys. **83**, 6494
17. Sides S. W., Rikvold P. A., Novotny M. A. (1999) Phys. Rev. E **59**, 2710
18. Tomé T., de Oliveira M. J. (1990) Phys. Rev. A **41**, 4251. These authors used the Kubo-Suzuki equation of motion corresponding to the Curie-Weiss form of the bistable free energy. However, equivalent results are obtained from a simple ϕ^4 potential.
19. Lo W. S., Pelcovits R. A. (1990) Phys. Rev. A **42**, 7471
20. Acharyya M., Chakrabarti B. (1995) Phys. Rev. B **52**, 6550
21. Acharyya M. (1997) Phys. Rev. E **56**, 1234
22. Acharyya M. (1997) Phys. Rev. E **56**, 2407
23. Acharyya M. (1998) Phys. Rev. E **58**, 179
24. Sides S. W., Rikvold P. A., Novotny M. A. (1998) Phys. Rev. Lett. **81**, 834
25. Buendía G. M., Machado E. (1998) Phys. Rev. E **58**, 1260
26. Acharyya M., Chakrabarti B. K. (1994) In: Stauffer, D. (Ed.) Annual Reviews of Computational Physics I. World Scientific, Singapore, 107
27. Chakrabarti B., Acharyya M. (1999) Rev. Mod. Phys. **71**, April
28. Binder K., Heermann D. W. (1997) Monte Carlo Simulation in Statistical Physics. An Introduction, 3rd edn. Springer, Berlin
29. Lubachevsky B. D. (1987) Complex Systems **1**, 1099
30. Lubachevsky B. D. (1988) J. Comput. Phys. **75**, 103
31. Korniss G., Brown G., Novotny M. A., Rikvold P. A. Preprint cond-mat/9803118, to appear in Computer Simulation Studies in Condensed Matter Physics XI, edited by D. P. Landau and H.-B. Schüttler. Springer, Berlin
32. Korniss G., Novotny M. A., Rikvold P. A. (1998) Preprint cond-mat/9812344, submitted to J. Comput. Phys.
33. Rikvold P. A., Tomita H., Miyashita S., Sides S. W. (1994) Phys. Rev. E **49**, 5080
34. Gunton J. D., San Miguel M., Sahni P. S. (1983) In: Domb C., Lebowitz J. L. (Eds.) Phase Transitions and Critical Phenomena Vol. 8. Academic, New York
35. Rikvold P. A., Gorman B. M. (1994) In: Stauffer, D. (Ed.) Annual Reviews of Computational Physics I. World Scientific, Singapore, 149
36. Avrami M. (1939) J. Chem. Phys. **7**, 1103; (1940) J. Chem. Phys. **8**, 212; (1941) J. Chem. Phys. **9**, 177
37. Ramos R. A., Rikvold P. A., Novotny M. A. (1999) Phys. Rev. B **59**, 1 April
38. Sides S. W., Rikvold P. A., Novotny M. A. (1998) Phys. Rev. E **57**, 6512
39. Binder K. (1990) In: Privman V. (Ed.) Finite-Size Scaling and Numerical Simulation of Statistical Systems. World Scientific, Singapore, 173
40. Landau D. P. (1976) Phys. Rev. B **13**, 2997; (1990) In: Privman V. (Ed.) Finite-Size Scaling and Numerical Simulation of Statistical Systems. World Scientific, Singapore, 225
41. Stauffer D., Aharony A. (1992) Introduction to Percolation Theory, 2nd edn. Taylor & Francis, London
42. Sides S. W., Rikvold P. A., Novotny M. A. (unpublished)
43. Kamieniarz G., Blöte H. (1993) J. Phys. A **26**, 201

Application of the 8-State Clock Model to FM/AFM/FM Trilayers

P. D. Scholten and V. Ranjbar

Department of Physics, Miami University,
Oxford, Ohio 45056, USA

Abstract. The phase diagrams of magnetic trilayers consisting of three or four AFM planes sandwiched between two layers of FM planes were studied by Monte Carlo simulation. A bcc lattice was used for the structure of the each system, and an 8-state clock model was used for the individual spins. The discrete nature of this model played a significant role at low temperatures, producing a series of first-order transitions.

Systems consisting of alternating layers of ferromagnetic (FM) and antiferromagnetic (AFM) films have been of considerable interest both experimentally and theoretically. Previous theoretical studies approximated the magnetization of each layer by a single spin and examined ground state configurations of systems composed of an infinite number of alternating layers of FM and AFM planes [1,2]. Other work used a mean field approximation to produce phase diagrams for temperatures greater than zero but with no external field [3]. The only previous Monte Carlo work done on such a FM/AFM multilayer system used an Ising model to represent the spins [4].

Our purpose was to use Monte Carlo simulations to map out the complete field-temperature phase diagrams for a FM/AFM/FM trilayer composed of two-dimensional spins. We chose an eight-state clock model for this work to allow the spins to rotate in two dimensions while at the same time avoiding the high computational cost of a continuous spin model.

The particular systems used consisted of one AFM layer of either three or four AFM planes sandwiched between two FM layers, each having three planes. These two cases will be designated as 3-3-3 and 3-4-3 structures, respectively. Each plane had LxL sites, and all planes were part of a bcc lattice. The data reported here are principally for L=10, although some data for L=14, 20, and 30 were also collected to study finite-size effects. Periodic boundary conditions were used in both directions parallel to the planes, but free surfaces were used in the direction perpendicular to the planes. Both nearest-neighbor (nn) and next-nearest-neighbor (nnn) interactions were included. For this study the magnitudes of all nn exchange interactions were set equal to one (Jnn=1). The nn coupling at the layer interfaces was antiferromagnetic. The ratio of nnn to nn exchange interactions was Jnnn/Jnn = $3\sqrt{3}/8$ to model a $1/r^3$ spatial dependence. Uniaxial anisotropy was not included. With the bcc structure all spins within a plane

Springer Proceedings in Physics, Volume 85
Computer Simulation Studies in Condensed-Matter Physics XII
Eds.: D.P. Landau, S.P. Lewis and H.-B. Schüttler

were nnn, and hence all spins in each plane were parallel to one another in the ground state. The most significant difference between the ground states of the 3-3-3 and 3-4-3 structures is that the spins in the two FM layers pointed in the same direction in the former case, but were antiparallel in the latter.

Two types of Monte Carlo runs were done to investigate the phase diagrams. A general idea of system behavior was obtained by sweeping either temperature or field. Locations of critical points were determined by locating the maxima in the system heat capacity and susceptibility. At least 105 MCSS were done at each value of kT/Jnn and H/Jnn. This method worked well at all but the lowest temperatures where hysteretic effects due to first-order transitions were strong.

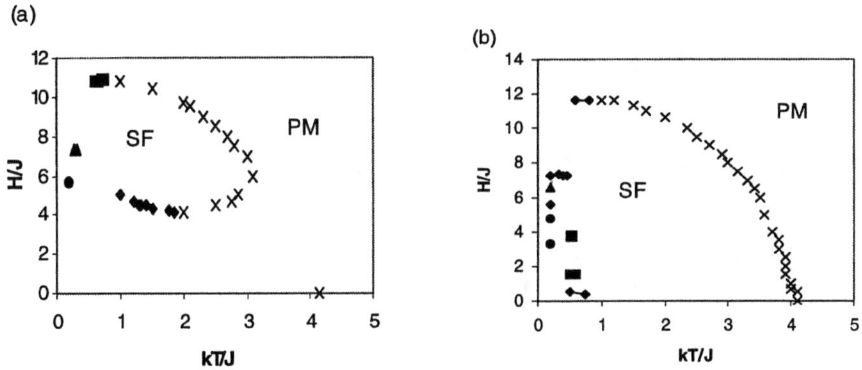

Fig. 1. Phase diagram for the trilayer systems: **(a)** 3-3-3 structure, **(b)** 3-4-3 structure. The antiferromagnetic state exists along the *H/J=0* line. The other observed phases are paramagnetic (PM) and spin flop (SF). The *crosses* denote continuous transitions; the *solid symbols* denote first-order transitions

To study first-order transitions in more detail it was necessary to do longer Monte Carlo runs (typically 107 MCSS) to construct distribution functions of the energy and the components of the magnetization for the system. Histogram methods were used to extend the distributions to nearby values of kT/Jnn and H/Jnn. This technique proved useful only very close to the temperatures of critical end points or tricritical points. At lower temperatures the transitions were so strongly first-order that adequate sampling could not be done, even for these small systems. Locations of first-order transitions at lower temperatures were estimated from observed hysteresis loops in the system order parameter.

The phase diagram of 3-3-3 structure is shown in Fig. 1 a. For H=0 the system undergoes a continuous transition from the AFM state to the paramagnetic (PM) state at $kT/Jnn = 4.05 \pm 0.03$. For fields greater than about H/Jnn=4 a spin flop (SF) phase appears. Transitions between the SF and PM phases are continuous for $kT/Jnn > 2$ at low fields and for $kT/Jnn >$

0.9 at high fields. At lower temperatures the transitions are first-order. The locations of the two tricritical points are kT/Jnn $= 1.88 \pm 0.05$, H/Jnn $=$ 4.15 ± 0.05 and kT/Jnn $= 0.80 \pm 0.05$, II/Jnn $= 10.90 \pm 0.05$. No transitions were found at any temperature in the region $0 <$ H/Jnn < 4.

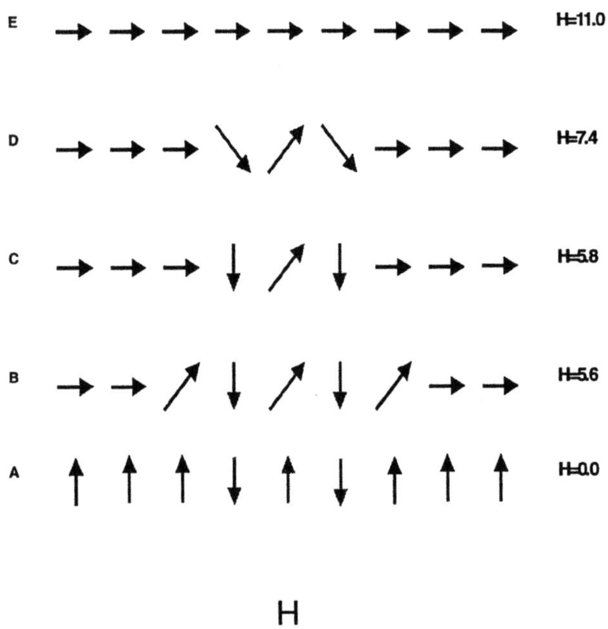

Fig. 2. Planar spin configurations for the 3-3-3 structure at kT/Jnn=0 and the values of H/Jnn at which they first appear

To study the low temperature states of the system in more detail we used a simple model wherein all spins in a plane were parallel and hence could be represented by a single spin. The nn and nnn interplanar couplings were rescaled to allow comparison with the Monte Carlo data. A thorough search was then done to find the spin configurations that minimized the system energy as the external field was increased. Fig. 2 shows these configurations for the 3-3-3 structure and the fields at which they first appear. The presence of such configurations are due, of course, to the discrete nature of the 8-state clock model. The long histogram Monte Carlo runs at low temperatures were able to find all of these configurations at fields close to those given by the single spin model. The two first-order lines separating the B and C and the C and D configurations are both inside the SF region and end in critical points at kT/Jnn $= 0.24 \pm 0.02$, H/Jnn $= 5.66 \pm 0.03$ and kT/Jnn $= 0.38 \pm 0.04$, H/Jnn $= 7.36 \pm 0.03$, respectively. The first-order lines separating the A and

B and the D and E configurations form parts of the SF/PM boundary as described above.

The phase diagram for the 3-4-3 structure (see Fig. 1 b) is qualitatively different in that the SF/PM boundary extends all the way down to H/Jnn=0. The low temperature region has more first-order transitions, indicating that there are more intermediate planar spin configurations between the AFM ground state and the PM state with all spins parallel. Again, the Monte Carlo simulations found all nine configurations indicated by the single spin model. Beyond the critical points within the SF phase the discrete nature of the model was not evident as the magnetization angles of the individual planes rotated smoothly as the field was increased. Work is now underway to explore the effects of including uniaxial anisotropy and of varying the spin-spin interactions.

References

1. Hinchey, L. L. and Mills, D. L. Phys. Rev. B33, 3329 (1986)
2. Genkin, G. M., Sapozhnikov, M. V., and Tokman, I. D. J. Mag. Mag. Mat. 132, 323 (1994)
3. Wang, X. Z. and Zhang, Z. J. Phys.: Condens. Matter 2, 5765 (1990)
4. Li J. and Landau, D. P. J. App. Phys. 69, 6158 (1991)

Monte Carlo Study of Critical Point Shifts in Thin Films

O. Dillmann[1], M. Müller[1], W. Janke[2], and K. Binder[1]

[1] Institut für Physik, Johannes Gutenberg-Universität, 55099 Mainz, Germany
[2] Institut für Theoretische Physik, Universität Leipzig, 04109 Leipzig, Germany

Abstract. We report preliminary results of Monte Carlo simulations of critical point shifts in thin slit-like capillaries. By making use of the isomorphism with an Ising model subject to bulk and surface fields and employing a multi-cluster update algorithm with ghost-spin term we obtain the coexistence curve and the behavior at the critical point for various film thicknesses D.

1 Introduction

Liquids in thin slit-like capillaries and liquid binary mixtures in thin films on surfaces are of growing importance for material sciences [1]. Common applications are, e.g., coating of surfaces and lubrication. But thin films are also of interest from a theoretical point of view [2,3]. Their critical behavior differs from that of the bulk, e.g., the critical point is shifted. Qualitatively thin films show a cross-over from two- to three-dimensional critical behavior with increasing thickness. Responsible for these modifications are both the finite thickness of the film and the effects of the surface interactions.

2 Model

For studying the properties of a thin liquid film, we consider a lattice gas confined into a small slit of thickness D between two walls. The corresponding Hamiltonian is of the following form

$$\mathcal{H}(\{c_i\}) = -\varphi \sum_{\langle i,j \rangle} c_i c_j - \epsilon_1 \sum_{i \in \text{surfaces}} c_i - \mu \sum_i c_i \,, \tag{1}$$

where c_i is the occupation variable (i.e., $c_i = 1$ if a site of the lattice is occupied, and $c_i = 0$ else). There is an attractive interaction $\varphi > 0$ between nearest-neighbor particles $\langle i,j \rangle$ in the $L \times L \times D$ system and an attractive binding energy of the walls $\epsilon_1 > 0$ whose range is restricted to the adjacent layers. The chemical potential is denoted by the variable μ.

If we describe the system state using a spin variable $S_i = 1 - 2c_i = \pm 1$, the lattice gas model translates into the Ising ferromagnet,

$$\mathcal{H}(\{c_i\}) = \mathcal{H}_0 + \mathcal{H}_{\text{Ising}}(\{S_i\}) \,, \tag{2}$$

Springer Proceedings in Physics, Volume 85
Computer Simulation Studies in Condensed-Matter Physics XII
Eds.: D.P. Landau, S.P. Lewis and H.-B. Schüttler

where $\mathcal{H}_0$ is a constant and

$$\mathcal{H}_{\text{Ising}}(\{S_i\}) = -J \sum_{\langle i,j \rangle} S_i S_j - H \sum_i S_i - H_1 \sum_{i \in \text{surfaces}} S_i \,. \tag{3}$$

The interaction between nearest neighbors is now governed by the exchange constant $J = \varphi/4$ and the influence of the walls is now described by a magnetic surface field $H_1 = -\epsilon_1/2 - J$. To the chemical potential corresponds a magnetic bulk field $H = -\mu/2 - zJ$, with z being the coordination number of the lattice.

3 Phase Coexistence Curve and Scaling Predictions

For examining the phase coexistence curve in the (T, H)-plane, we start with a film of infinite thickness D ($D \to \infty$). In this case the magnetic surface field H_1 can be neglected because of its restricted range, and the phase coexistence between all spins S_i up and all spins down can only occur for a vanishing magnetic bulk field $H = 0$ for symmetry reasons. This holds for all temperatures T below the critical temperature T_c and therefore the phase coexistence curve is given by the part of the T-axis which is located below T_c, see Fig. 1.

For finite D the situation is different. In the limiting case of $T = 0$ equating the energies of phases with opposite spin orientation yields

$$H_{\text{coex}}(D, T = 0) = -2H_1/D \,, \tag{4}$$

but $H_{\text{coex}}(D, T)$ depends non-trivially on the temperature and ends in a critical point that is shifted compared to the bulk critical point in T and H direction. See Fig. 1 for a sketch of the coexistence curve for finite D.

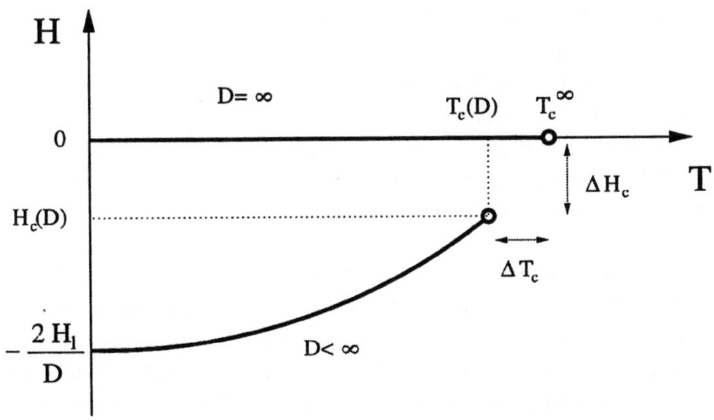

Fig. 1. Qualitative sketch of the phase coexistence curve

For describing the shift of the critical point in thin films, Fisher and Nakanishi [4] derived the scaling relation

$$f(D;T,H,H_1) \approx |t|^{2-\alpha}\, W(D|t|^{\nu}; H/|t|^{\Delta}, H_1/|t|^{\Delta_1}), \tag{5}$$

where $t \equiv |1 - T/T_c|$, α and ν are the standard critical exponents of the specific heat and correlation length, respectively, $\Delta = \beta + \gamma$ is the bulk gap exponent, and Δ_1 is the special surface field exponent. In mean field theory these exponents take the values $\alpha = 0$, $\nu = 1/2$, $\Delta = 3/2$, and $\Delta_1 = 1/2$.

By choosing the scaling parameters $x = D|t|^{\nu}$ and $w = H_1 D^{\Delta_1/\nu}$, eq. (5) becomes

$$f \approx |t|^{2-\alpha}\, W\left(x; \frac{HD^{\Delta/\nu}}{x^{\Delta/\nu}}, \frac{w}{x^{\Delta_1/\nu}}\right). \tag{6}$$

In particular we can read off from eq. (6) the shifts of the critical point [4],

$$\Delta T_c = -B_T D^{-1/\nu}\, X_c(C_T H_1 D^{\Delta_1/\nu}),$$
$$\Delta H_c = -B_H D^{-\Delta/\nu}\, Y_c(C_H H_1 D^{\Delta_1/\nu}),$$

where B_T, C_T, B_H, and C_H are constants.

4 Monte Carlo Procedures and Results

A Monte Carlo (MC) simulation was carried out, simulating a $L \times L \times D$ lattice with periodic boundary conditions in the $L \times L$-planes parallel to the walls and two free $L \times L$-surfaces. For updating the lattice a multi-cluster algorithm (Swendsen–Wang) [5] was chosen. To account for the magnetic fields the standard algorithm was modified by introducing a so-called *ghost spin* [6]. In a first step, the lattice was divided into clusters by connecting neighboring spins with an "active" bond with a probability $p = 1 - e^{-2\beta J}$ if they are in the same state. Here $\beta \equiv 1/T$ denotes the inverse temperature. If not, there is no connecting bond. Then a ghost spin pointing in the direction of the total magnetic field h (accounting for both H and H_1) was introduced and each single spin of the lattice pointing in the same direction was connected to it by a ghost bond with a certain probability $p_g = 1 - e^{-2\beta|h|}$. Finally the lattice was updated by choosing randomly a new spin orientation for each cluster not connected to the ghost spin, i.e. a cluster connected to the ghost spin never flips.

For the determination of the critical point shift for a certain thickness D and a fixed surface field $H_1 = -0.015$ we took the following steps. First we generated MC time-series data for different system sizes $L = 16, 32, 64, 128, \ldots$ at different temperatures and bulk fields. Then we obtained the coexistence curve in the (T, H)-plane using reweighting techniques. Afterwards the location of $T_c(D)$ on the coexistence curve is determined by the intersection of the cumulant $U_{4/2} = 1 - \langle m^4 \rangle / \left(3\langle m^2 \rangle^2\right)$ for different L, see Figs. 2 and 3.

To confirm that the asymptotic limit is reached we tried to map the probability function $P(m)$ onto the scaling function corresponding to the 2D Ising universality class. The quality of such a mapping can be inspected in Fig. 4.

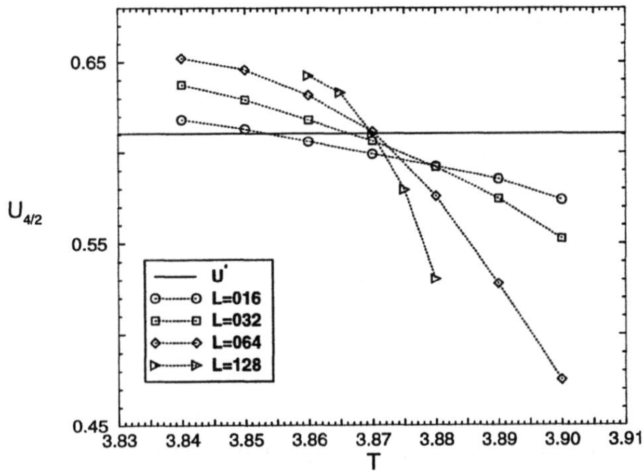

Fig. 2. Cumulant intersection for systems with different L at the critical point for film thickness $D = 4$. A surface field of $H_1 = -0.015$ is applied. U^* denotes the universal value of the intersection for the 2D Ising model

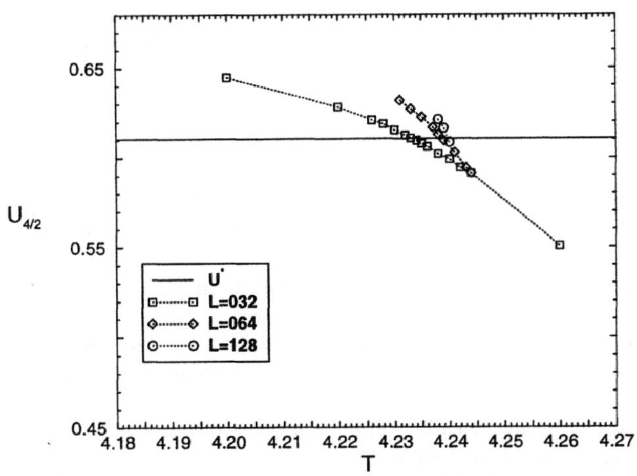

Fig. 3. Same as in Fig. 2 for film thickness $D = 8$

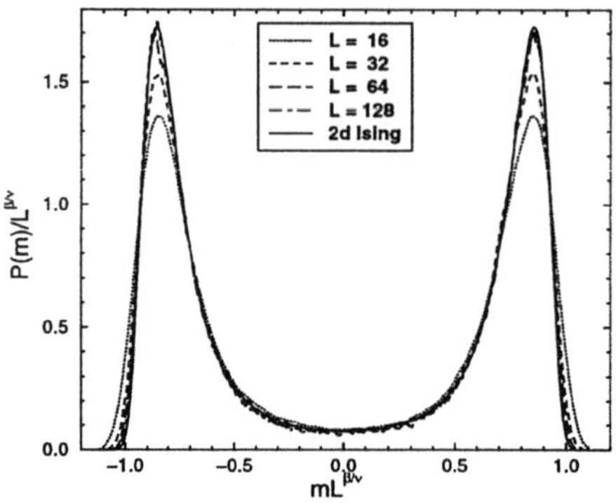

Fig. 4. Finite-size scaling plot of the magnetization m for a film of thickness $D = 4$ with surface field $H_1 = -0.015$ at phase coexistence

5 Summary and Outlook

We are now in the position of having all necessary tools for the examination of the critical point in our model, but also the complete coexistence line is of physical interest. So we are currently adapting multicanonical techniques to this problem. The ultimate goal of this study, however, is to investigate the Fisher–Nakanishi scaling hypothesis. To this end we are currently running simulations for further film thicknesses.

Acknowledgement

OD gratefully acknowledges support from the Deutsche Forschungsgemeinschaft (DFG) (No. BI 314/16).

References

1. Croxton C. A. (ed.) (1986) Fluid Interfacial Phenomena. Wiley, New York
Gregg S. J., Sing K. S. W. (1982) Adsorption, Surface Area and Porosity. Academic, New York
2. Binder K., Landau D. P. (1992) Capillary Condensation in the Lattice Gas Model: A Monte Carlo Study. J Chem Phys 96:1444–1454
3. Binder K., Ferrenberg A. M., Landau D. P. (1994) Wetting and Capillary Condensation of Lattice Gases in Thin Film Geometry. Ber Bunsenges Phys Chem 98:340–345
4. Fisher M. E., Nakanishi H. (1981) Scaling Theory for the Criticality of Fluids Between Plates. J Chem Phys 75:5857–5863
5. Swendsen R. H., Wang J.-S. (1987) Nonuniversal Critical Dynamics in Monte Carlo Simulations. Phys Rev Lett 58:86–88
6. Dotsenko V. I., Selke W., Talapov A. L. (1991) Cluster Monte Carlo Algorithms for Random Ising Models. Physica A 170:278–281

Incipient Spanning Clusters in Square and Cubic Percolation

L. N. Shchur

Landau Institute for Theoretical Physics, 142432 Chernogolovka, Russia

Abstract. The analysis of extensive numerical data for the percolation probabilities of incipient spanning clusters in two dimensional percolation at criticality are presented. We developed an effective code for the single-scan version of the Hoshen-Kopelman algorithm. We measured the probabilities on the square lattice forming samples of rectangular strips with widths from 8 to 256 sites and lengths up to 3200 sites. At total of more than 10^{15} random numbers are generated for the sampling procedure. Our data confirm the proposed exact formulaes for the probability exponents conjectured recently on the base of 2D conformal field theory. Some preliminary results for 3D percolation are also discussed.

1 Introduction

Percolation is a quite common phenomena in nature. Among known examples there are epidemic diseases of garden trees, forest fires, the process of espresso preparation, Ising spins at criticality, etc. [1]. Ising spins are correlated with probability $p = 1 - \exp(-2J/k_BT)$ (J is coupling constant, k_B - Boltzmann factor and T - temperature) and clusters of parallel spins percolate at the critical temperature T_c [2].

Percolation is the simplest example of a critical phenomenon. Scaling relations for infinite lattices were developed in the same way as for the Ising model [3]. Mapping of the percolation model onto a $q = 1$ state Potts model [2] gave a way to predict critical exponents for percolation in two dimensions using techniques of Conformal Field Theory (CFT) or of Coulomb gas representation.

In the last few years, new insights into the properties of percolation clusters on infinite lattices at the critical concentration have been developed. First of all, Langlands and coauthors found numerically [4] that the probability of a cluster spanning the lattice horizontally is an universal function of the aspect ratio only, and does not depend on the lattice symmetry or on the type of percolation (see also [5]).

Second, Aizenman proved that there could be more than one incipient spanning cluster in two-dimensional critical percolation [6]. His results were in contradiction with a wide-spread belief that the percolation cluster is unique (see, for example [7]). Numerical results for percolation in a strip demonstrated that there could be more than one cluster spanning the short

Springer Proceedings in Physics, Volume 85
Computer Simulation Studies in Condensed-Matter Physics XII
Eds.: D.P. Landau, S.P. Lewis and H.-B. Schüttler

direction in a strip [8]. Later Aizenman's proposal was supported numerically for the case of bond percolation [9] (see also [10]).

Finally, Cardy conjectured an exact form for the probability that k clusters span a large strip or open cylinder of aspect ratio r, in the limit when k is large [11]. His result is in a complete agreement with the Aizenman theorem.

The assumptions of conformal field theory are not rigorously established, and it remains important to perform precise numerical tests of the theory. I present here an extensive numerical verification of Cardy's proposal. The numerical data were obtained in collaboration with Sergey Kosyakov [9,12,31].

In Section 2 some known facts on percolation probabilities are shortly reviewed. Spanning probability are discussed in Section 3. Most recent findings on the multiplicity of incipient spanning clusters (ISC), are discussed in Section 4, where our previously published results for the ISC probabilities in bond percolation are compared with the proposed exact result. The computational method is described in the Section 5. New results for the finite size corrections of spanning probabilities in a strip geometry with up to 819200 sites are presented in Sections 6 and 7 for open boundaries and cylinder geometry. Preliminary results for the spanning probabilities in a simple cubic critical percolation with the samples with up to 1900544 sites are presented in Section 8. All averages were done using 10^8 strips in all geometries investigated. Discussion and acknowledgements finish the paper.

2 Percolation

The simplest model of percolation is bond percolation on the square lattice. Let us consider a two dimensional (2D) rectangle with an infinite number of sites L in both (vertical and horizontal) directions. The probability for the bond to be occupied (or placed, or black) is equal to p, $0 \le p \le 1$. Consequently, the probability for the bond to be open (or removed, or white) is equal to $q = 1 - p$. If all bonds are closed (that is $p = 1$) the left side is connected with the right one by closed bonds. That is closed, paths from the left boundary to the right boundary exist. Clearly, if all bonds are open ones ($p = 0$) there is no way to go by closed bonds through the rectangle. It appears that there is some value of the probability p_c dividing these two possibilities. The value of critical probability p_c is equal to $1/2$ for bond percolation in two dimensions [13,14]. This can be seen from the self-duality of the 2D square lattice.

The dual lattice is constructed placing open bonds across the closed bonds on the prime lattice, and by drawing closed dual bonds across the open prime bonds. An example configuration is plotted in Fig. 1. It is clear from the construction that the number of closed prime bonds is equal to the number of open dual bonds. If p is the probability that a bond is closed on the prime lattice, then $q = 1-p$ is the probability that the chosen bond be closed on the dual lattice. For a given probability p of closed bonds, let us denote by $P_h(p)$

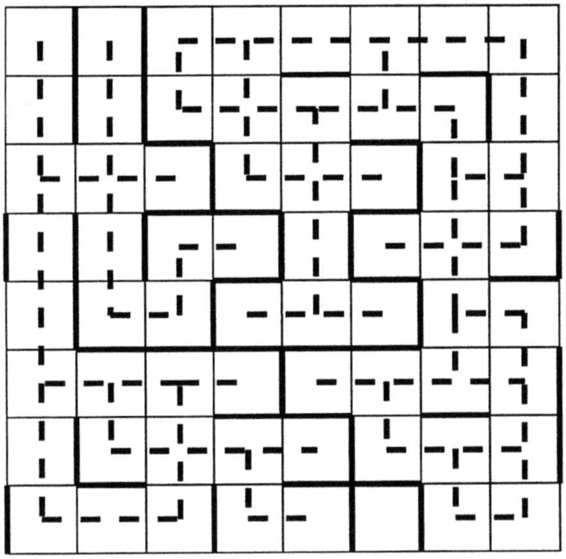

Fig. 1. Example of a prime lattice and its dual. Closed bonds of the prime lattice are denoted by bold full lines, and those of the dual lattice by bold dashed lines. Open prime bonds are plotted with thin full lines. Open dual bonds are not shown

the probability that there is a path of connected closed bonds connecting left and right sides of the prime lattice. That is $P_h(p)$ is the probability of a cluster spanning the lattice horizontally. By $P_v^*(q)$ we will denote probability that a cluster of closed bonds spans the dual lattice vertically for a given probability q of closed dual bonds. Two relations follows

$$P_h(p) + P_v^*(q) = 1$$
$$P_v(p) + P_h^*(q) = 1. \tag{1}$$

The prime and dual lattices are identical in this construction. Therefore, the percolation threshold should be at $p_c = q_c$, i.e. $p_c = 1/2$. A strong proof of that statement is the main subject of Kesten's book [14].

For the finite system with square size $L \times L$ the probability $P_h(p; L)$ of a horizontal crossing is shown on Fig. 2 as a function of p for several L. The slope of this curve grows with the lattice size L as $L^{1/\nu}$, ν being the correlation length exponent [3]. Is is clearly visible that $P_h(p; L)$ approaches a step function as L tends to infinity. Indeed, there are strong mathematical results [14,15] supporting this picture.

What is the value of the spanning probability exactly at the percolation threshold p_c? It was believed for a long time that this value was equal to the value of the percolation threshold [3]. Indeed, this is true for the bond

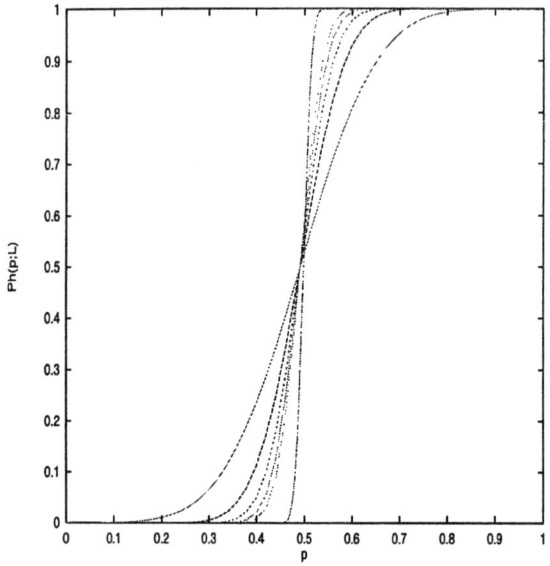

Fig. 2. Spanning probability $\pi_h(p, L)$ as function of bond occupation probability p for several values of lattice sizes $L = 10, 20, 30, 40, 50$ and 200. The larger lattice size the larger derivative of spanning probability

percolation on the square lattice. From relations (1) for the bond percolation on the infinite square lattice it is clear that

$$P_h(p_c) = P_v^*(q_c) = P_v(p_c) = P_h^*(q_c) = 1/2. \tag{2}$$

At the same time, the real space renormalization analysis [16] are based on the relation $P_h(p) = p$ for the site percolation. In the last case $p_c \approx 0.592746$, which is fairly close to 0.5 and reasonable results for the critical exponents were computed. However it was recognized later that the value of spanning probability $P_h(p_c)$ does not depend on the type of percolation as the percolation threshold does. Moreover it turns out that this value is invariant and depends only on the aspect ratio of the rectangle [4,17,19].

3 Incipient Spanning Probability

All results described in this section as well as in the following sections are obtained for critical percolation. The probability p for the bond (or site) to be closed is equal to the corresponding critical value p_c throughout the remaining text. The value of p_c is not universal and depends on the dimensionality, lattice type (square, honeycomb, etc.) and percolation type (site, bond, etc.).

The spanning probability appears to be universal and invariant under shape transformations of the rectangle, also depends on the boundary conditions [4,17,19] (see also the mini review by Stauffer [20]).

Let us consider a strip with vertical size $L_v = L$ and horizontal size $L_h = rL$, where $r = L_h/L_v$ is the aspect ratio. Incipient spanning clusters connect two segments on the boundary of a macroscopically large strip. Langlands et al. [4,19] found numerically evidence that the probability that a such cluster exists is invariant under conformal transformation. Cardy conjectured an exact formula for this probability [17] in terms of hypergeometric functions

$$P_h(\eta) = \frac{3\Gamma(\frac{2}{3})}{\Gamma(\frac{1}{3})^2}\eta^{\frac{1}{3}}\ {}_2F_1(\frac{1}{3},\frac{2}{3},\frac{4}{3};\eta),$$

where Γ and ${}_2F_1$ are the gamma function and hypergeometric function respectively and $\eta = ((1-k)/(1+k))^2$, where k defines the aspect ratio r of the rectangle $r = K(1-k^2)/2K(k^2)$ as the ratio of two complete elliptic integrals $K(u)$. The modulus k is associated with the positions $(-1/k, -1, 1, 1/k)$ on the real axes mapped under a Schwartz-Christoffel transformation to the vertices x_i of rectangle. Thus, the probability P_h is just the probability that there are closed paths (Incipient Spanning Clusters) which connect the interval $(x1, x2)$ with the opposite one $(x3, x4)$. This conjecture of Cardy was confirmed numerically by Langlands, et al. [19].

A more practical form of spanning probability was developed by Ziff as series expansion in powers of $\exp(-2\pi\, r)$ [21]

$$P_h(r) = \quad 2^{\frac{4}{3}} c\,(e^{-\pi r/3} - \tfrac{4}{7}e^{-7\pi r/3} + \tfrac{2}{13}e^{-13\pi r/3}...),\quad r \geq 1 \tag{3}$$

$$= 1 - 2^{\frac{4}{3}} c\,(e^{-\pi/3r} - \tfrac{4}{7}e^{-7\pi/3r} + \tfrac{2}{13}e^{-13\pi/3r}...),\ r \leq 1 \tag{4}$$

where $c \equiv \Gamma(\frac{2}{3})/\Gamma(\frac{1}{3})^2 \approx 0.566046680....$ Note, the corrections $\exp(-2\pi\, r)$ to the leading behaviour $\exp(-2\pi\, r/3)$ decay very fast with increasing r.

The probability that at least one cluster spans the square with open boundaries in both directions is equal to 1/2. This value could be obtained from (3) or (4) at $r = 1$.

In the case of cylinder the spanning probability along the cylinder is not known exactly. Simulations by Hovi and Aharony [22] give the value $P_h(1) = 0.6366(8)$ for the aspect ratio $r = 1$.

In a forthcoming paper [31] we compute the spanning probabilities $P_h(r)$ for the aspect ratio r in the interval $0.25 \leq r \leq 25$ for cylindrical geometry as well in the interval $0.25 \leq r \leq 12.5$ in open geometry with the step $\Delta r = 0.25$. The quality of these results will be discussed in the following sections.

4 Coexistence of Incipient Spanning Clusters in 2D

In the following sections we will also compute the probabilities that at least two, and even more clusters simultaneously span the lattice from left to right.

It was a common belief until recently that percolation clusters are unique on the 2D lattice. Aizenman proposed quite recently in his lecture at the Statistical Physics 19 Conference in China [23] that the number of Incipient

Spanning Clusters (ISC) in 2D can be larger than one. Later he proved [6] that the probability $P(k,r;L)$ that at least k ISCs span horizontally (that is along L_h) the strip $(L_h, L_v) = (rL, r)$ with width L_v and horizontal length L_h bounded

$$Ae^{-\alpha k^2 r} \le P(k,r;L) \le e^{-\alpha' k^2 r}, \tag{5}$$

where α and α' are different positive constants expected to merge for an infinite lattice [6]. Indications of the existence of simultaneous clusters in two-dimensional critical percolation in the limit of infinite lattices were found in computer simulations by Sen[10] for site percolation on square lattices with helical boundary conditions and in a short strips by Hu and Lin [8] using their Monte Carlo histogram method.

We checked numerically the number of spanning clusters in the critical bond percolation model on two-dimensional square lattices [9]. We have determined the numerical values of probabilities

$$P(k,1) = \lim_{L\to\infty} P(k,1;L)$$

for $k = 1, 2$, and 3 by means of finite-size scaling. The values are given in Table 1 for the cases of free boundary conditions and periodic boundary conditions in the vertical direction (i.e. perpendicular to the spanning direction).

Table 1. Probability to have at least k horizontally spanning clusters in square with free boundaries (FBC) and periodic boundaries (PBC). Bond critical percolation

n	$P(k,1)$ - FBC	$P(k,1)$ - PBC
1	0.50002(2)	0.6365(1)
2	0.00658(3)	0.0020(4)
3	0.00000148(21)	0.00000014(5)

Actually, we use fully symmetric lattices, which are self dual even at finite sizes. Therefore, relations (2) holds for our finite lattices (for details, see [9]) and $P(1,1;L) \equiv 1$. In fact, the data in Table 2 support our observation. Finite size corrections for the probabilities of at least 2 incipient spanning clusters appear to be of the form $P(2,1;L) = P(2,1) + c/L^2$, with some constant c, i.e. proportional to the inverse square lattice size L. We found that the same finite size dependence holds also for $P(3,1;L)$ [9].

The values of probabilities in Table 1 tell us, that at criticality there is percolation in one direction with the probability equal to 1/2. Among such samples and in the case of free boundaries, in about 6 samples in 1000 there are two different clusters percolated horizontally. Finally, 1 sample in 1000000 (one million) contain at least 3 clusters spanning lattice in the given direction. The corresponding probabilities for $k > 1$ are less in the case of periodic boundary conditions, that is for cylindrical geometry.

Table 2. Probabilities $P(k, 1; L)$ of at least k incipient spanning clusters on critical bond square lattices with size (L, L) and free boundaries. Note, that the values of $P(2, 1; L)$ are multiplyed by a factor 10^3 and those of $P(3, 1; L)$ by a factor 10^6. For each lattice size L the first row is the probability of horizontal spanning and the second one is the probability of vertical spanning

L	$P(1,1;L)$	$P(2,1;L) \cdot 10^3$	$P(3,1;L) \cdot 10^6$
8	0.50005(5)	7.657(8)	3.40(15)
	0.50003(4)	7.660(8)	3.98(21)
12	0.50002(5)	7.084(9)	2.57(14)
	0.49995(5)	7.070(8)	2.10(13)
16	0.50003(7)	6.855(9)	1.97(17)
	0.50002(6)	6.843(8)	1.79(19)
20	0.49990(6)	6.742(8)	1.95(14)
	0.50008(5)	6.745(9)	1.72913)
30	0.49999(4)	6.650(8)	1.52(14)
	0.49996(5)	6.653(7)	1.52(12)
32	0.49999(5)	6.648(8)	1.73(12)
	0.50008(7)	6.642(8)	1.56(11)
64	0.49992(9)	6.597(9)	1.33(13)
	0.49999(6)	6.602(8)	1.51(14)

The data presented in the Table 1 appear in good agreement with the exact form of the probabilities $P(k, r)$ conjectured by Cardy using methods of conformal field theory [11]. He extend the arguments of his early paper [17] and determined the exact behaviour in the limit of infinite lattices $L \to \infty$ for the spanning probabilities $P(k, r; L)$ for large aspect ratios r in the case of free boundaries

$$\lim_{L\to\infty} \ln P_{FBC}(k, r; L) \propto -\frac{2\pi}{3} k(k - \frac{1}{2})\, r, \tag{6}$$

as $r \to \infty$ for any k.

Analogously, for periodic boundaries in the vertical direction (spanning along a cylinder) he found

$$\lim_{L\to\infty} \ln P_{PBC}(k, r; L) \propto -\frac{2\pi}{3} (k^2 - \frac{1}{4})\, r, \tag{7}$$

for $k \geq 2$ and $r \to \infty$, and with the different exponent for $k = 1$

$$\lim_{L\to\infty} \ln P_{PBC}(1, r; L) \propto -\frac{5\pi}{24}\, r, \tag{8}$$

in the limit of large aspect ratio $r \to \infty$.

From the results (6-7) one could calculate ratio

$$\frac{\ln P_{FBC}(k,1)}{\ln P_{PBC}(k,1)} = \frac{k(k-\frac{1}{2})}{k^2-\frac{1}{4}},$$

which is $\frac{4}{5} = 0.8$ for $k = 2$ and $\frac{6}{7} \approx .857$ for $k = 3$. Data from the Table 1 give us the values 0.808(10) and 0.857(20), respectively. This close agreement with the asymptotic form for $r \to \infty$ may be explained by the observation [11] that the corrections to the leading behaviour are of order $\exp(-2\pi r)$ (compare with expansion (4)).

The detailed numerical check of Cardy's results (6-8) is given in the remaining sections.

5 Computational Method

Our simulations can be divided into three major parts:

Sample choosing. There are two ways to generate a sample. The first one, is with the number of closed bonds (or sites) fixed, i.e. we choose the sample from a *canonical ensemble.* This method is especially convenient for the bond square percolation, when fully self-dual lattices exist even for the finite system size [9]. This eliminates not only boundary effects, but also 'fluctuations' of the concentration.
The second way, is with a fixed probability for a given site (bond) to be closed, i.e. sampling from a *grand canonical ensemble.* This method is more convenient for the Ziff hull method, for the transfer-matrix method and for the Hoshen-Kopelman algorithm we use in this work.
It is known that both methods lead to the same results in the 'thermodynamic' limit of infinite lattices [24]. It should be noted, that the computed averages coincide for both samples, whereas corresponding dispersions are quite different, as was found recently by Vasilyev and author [25].
To generate samples, we generate random numbers as result of XOR operations (eXclusive OR) applied to the output of two shift registers with large length. Usually, we use the pairs of legs (p,q) with $(9689, 471)$ and $(4423, 1393)$, though some other combinations from [18] could be used as well. This give us an enormously large period of random numbers and a lack of acting correlations. Details could be found in papers [26–29] and references therein.

Decomposition to clusters. We use the Hoshen-Kopelman algorithm [30] in their original "single scan version". That is we don't hold the whole lattice in the memory but only the boundary column we start from, the current column and the next column. In this way, the algorithm can be named as a "dimensionality reduction" Hoshen-Kopelman algorithm. This give us the possibility to fit all data together with the instruction code within the 4 MB cache memory of Alpha workstations. So, we avoid the slowest feature of modern computers, i.e. the slowing down of data flow from CPU registers to RAM and back, typical for general simulations. The resulting speed up of simulations is a factor of about 4.

Check of spanning. We check spanning with the step Δr in aspect ratio equal to 1/4 (and even $\Delta r = 1/8$ for the largest width $L = 256$). Namely, the sample generation goes from right to left together with the Hoshen-Kopelman labeling. After the successful generation of $L/\Delta r = L/4$ sequential columns, we compare the cluster labels on the start column with the cluster labels of the current column. The resulting information is the indicator functions of the aspect ratio $I_i(k,r)$, where k denotes the number of independent spanning clusters and i is the sample number. The value of this function at a fixed value of variable r is equal to 1 (there is spanning of k clusters) or 0 (there is no spanning of k clusters). This information is added to an output file. After 10^6 scans of the strip, the file is written onto the disk and a new file is opened. The process was repeated 100 times. So, each of 100 files contain the probabilities $\pi_h(k,r;L) = 10^{-6}\sum_{i=1}^{10^6} I_i(k,r)$ for the fixed value of strip width L. Error bars are computed over 100 such values of probabilities. It should be noted, that the probability $\pi_h(k,r;L)$ is the probability of exactly k clusters, and we have to compute $P(k,r;L) = \sum_{j=k}^{\infty} \pi_h(j,r;L)$ (i.e. sum from k to ∞) to get spanning probabilities [9] as defined in Eqs. (3,6-8).

This ends the 'measurement' of probabilities.

The detailed discussion of the computational methods will be published elsewhere [31].

The remaining part of the lecture is devoted to the analysis of the resulting data $P(k,r;L)$ [9,31].

6 Finite Size Corrections of Spanning Probabilities. Free Boundaries

In this section we present an analysis of the probability $P(k,r;L)$ that at least k clusters span an open strip of width L at least to a distance r. Free boundary conditions are used throughout this section. An analysis of the probabilities with cylindrical geometry will be given in the next section.

We generate 10^8 different strips of width $L_v = L =$8, 12, 16, 20, 24, 28, 32, 48, 64, 128, 256 sites growing up to the length of $L_h = rL = 32000$ sites. For the site percolation at $p_c = 0.592746$ considered here about 10^{15} random numbers were generated using 'XOR' combination of the output of two shift registers with the large lags. For some cases we repeat simulations with the linear congruential `rand48` generator (compare with [4]). Comparison did not demonstrate any evidence of systematic errors.

As results of the simulations, we have data for $P(k,r;L)$. Our next task is to determine the ISC exponents. For that, we have to take $\lim_{L\to\infty}$ and determine the slope of $\ln P(k,r)$ as a function of the aspect ratio r. This is the first method. Another one, is to determine the slope of $\ln P(k, r; L)$ and then take the limit of large strip widths $L \to \infty$.

Theoretically, both methods should give the same result. In practice the two methods gave slightly different values which should be comparable within error bars. Coincidence of both results could be a good test on the data accuracy. Here we will demonstrate how both methods works for the spanning probability $P(1, r; L)$ for which we have exact result (3) and then we will apply both methods to the probabilities of multiple ISCs, for which only results for very large aspect ratios r are known (6-8).

6.1 First Method

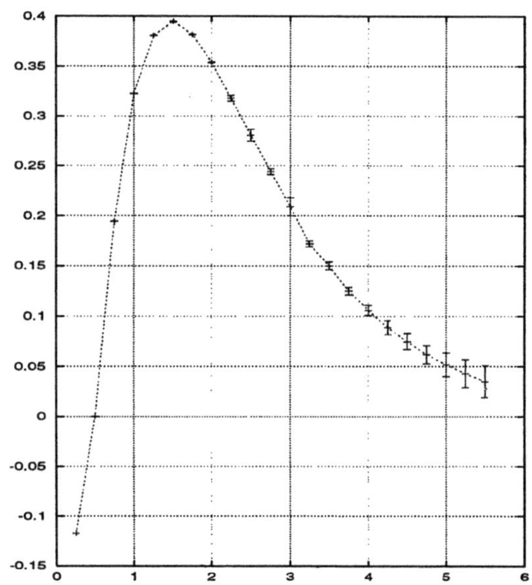

Fig. 3. Coefficient $b_0(r)$ in the fit to the spanning probability in 2D site percolation as the function of the aspect ratio r. See, Eqs. 10

First, we have to develop finite size corrections of probabilities. It is known from numerics, that the spanning probability for the site percolation on square lattices with aspect ratio $r = 1$ behaves at the critical point as [5]

$$P(1, 1; L) \approx P(1, 1) + \frac{b_0}{L + L_0} \tag{9}$$

with some constants $b_0 = 0.319$ and $L_0 = 1.6$.

Our data are in agreement with this result. We found that Ziff's proposal could be generalized [31] for any aspect ratio r

$$P(1,r;L) \approx P(1,r) + \frac{b_0(r)}{L + L_0(r)} \tag{10}$$

where $b_0(r)$ and $L_0(r)$ are now some functions of the aspect ratio r. $L_0(r)$ is a monotonic function vanishing very rapidly and can be set to zero for the aspect ratio $r \gg 2.5$. Function $b_0(r)$ is shown on Fig 3. It is interesting that the spanning probability for the site percolation does not depend on the lattice size for the aspect ratio close to 1/2. We could suggest that this aspect ratio $r = 1/2$ be used for the computation of percolation probability p_c with a higher accuracy than the usual simulations at $r = 1$.

Our resulting function $P(1,r)$ coincides with the exact one (3) within error bars. The linear fit to the logarithm of the resulting function $\ln P(1,r)$ in the interval $1.75 \le r \le 5.5$ gives the slop $\epsilon_1 = -1.0476(3)$ which is very close to Cardy's theoretical prediction $-\pi/3 = -1.047197$. To get an idea of the accuracy, the linear fit to Ziff's approximation in the same interval was done and yielded the slope $-1.0473(1)$. So, our first method produces quite accurate data.

6.2 Second Method

Let us now compute first the slope $\epsilon_1(L)$ of the curves $\ln P(1,r;L)$ with respect to the aspect ratio r for each strip width L, i.e. $\epsilon_1(L)$ and then take the limit

$$\lim_{L\to\infty} \epsilon_1(L) \to \epsilon_1 .$$

Values of $\epsilon_1(L)$ are given in the Table 3. A fit to the data in the table give

$$\epsilon_1(L) = \epsilon_1 + \frac{A}{L + L_0} = -1.04703(2) + \frac{1.3416(1)}{L + 1.7558(1)}, \tag{11}$$

with the value of ϵ_1 equal to the exact one in 4 digits.

It is interesting that the fit of values $\epsilon_1(L)$ with some exponent θ, i.e. with $\epsilon_1(L) = \epsilon_1 + A/L^\theta$ yield the result $\epsilon_1(L) = -1.05093(1) + 0.8365(2)/L^{0.8522(1)}$ with too large value of ϵ_1 and the exponent $\theta = 0.8522$ close to the irrelevant exponent, introduced by Stauffer (see, f.e. [3]) and included in the corrections to the scaling in the detailed analysis of invariance in two dimensional percolation [22]. Our data does not support this exponent but rather the expansion in powers of $1/L$ which could be effectively written as the term $A/(L + L_0)$. Such a term was used first in a paper by Levinstein, et al. (see, f.e. [32])[1] in their first-time computation of the correlation length exponent

[1] We are glad to acknowledge discussion of this point with E.I. Rashba and A.L. Efros, who stressed that they was introduced this term "to make the line as straight as possible".

and successfully used later in the context of spanning probabilities by Ziff [5] in his computation of the percolation threshold value $p_c = 0.592746$ for the site percolation, which is the most precise value (see also [33]). We found this form to be very helpful in the data analysis.

Table 3. Slope $\epsilon_1(L)$ of the $\ln P(1, r; L)$ as a function of the aspect ratio $r = L_h/L_v$ for several strip widths $L_v = L$. Slopes are computed in the range $1.75 \leq r \leq 5.5$

L	$\epsilon_1(L)$	σ
8	-0.909926	.0001
12	-0.949355	.00003
16	-0.971482	.0001
20	-0.985734	.0001
24	-0.995344	.00005
28	-1.001989	.00008
32	-1.007151	.0001
48	-1.019926	.00004
64	-1.026430	.00007
128	-1.036904	.00007
∞	-1.047032	.00002

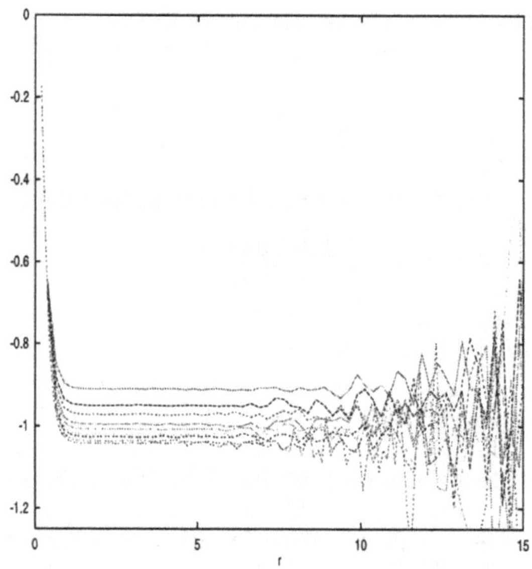

Fig. 4. Derivative of the logarithm of the spanning probability $P(1, r; L)$ as a function of the aspect ratio r for the strip width $L = 8, 12, 16, 24, 32, 48, 64, 128, 256$ from top to bottom. Site percolation on a square open strip

The second method could be modified in the following way. Taking the derivative of $\ln P(k,r)$ with respect to aspect ratio r we get

$$\frac{d}{dr}\ln P(k,r) \propto -\frac{2\pi}{3}k(k-\frac{1}{2}) = \epsilon_k. \tag{12}$$

This give us the idea the to compute $\epsilon_k(r,L)$ taking the numerical derivative of our data $\ln P(k,r;L)$. The result is shown on Fig. 4 for site percolation in a strip with several widths L. The slope of $\ln P(1,r;L)$ is constant in some region of r. Deviation below the left-hand value $r_{min} \approx 1.5$ is due to the corrections $\exp(-2\pi r)$ to the leading behavior as seen from Ziff's approximation (4) and becomes negligibly small starting from $r = 1.75$. Large fluctuations for aspect ratios larger than $r_{max} \approx 5.5$ is due to the small values of probabilities $P(1,r;L)$ for large r and, therefore, error bars become of the size of the fluctuations (not shown on the Fig. 4 for the reason of picture clarity). The Fig. 5 shows a magnified picture with the same derivative for the values of the aspect ratio $1.75 \le r \le 5.5$ together with the error bars.

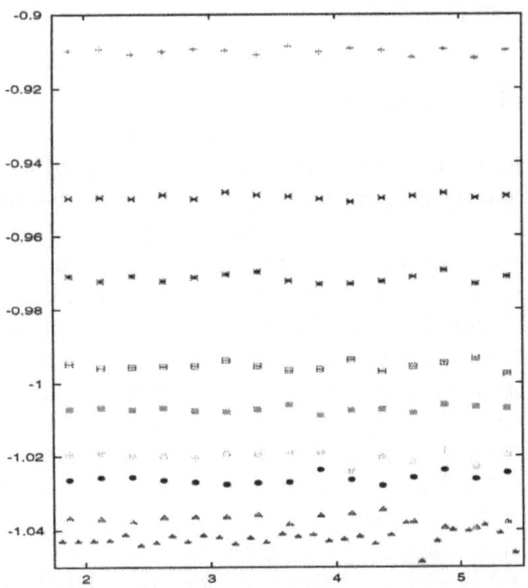

Fig. 5. The same as Fig.4 for the aspect ratio in the region $1.75 \le r \le 5.5$. Derivative of the logarithm of the spanning probability $P(1,r;L)$ for strip widths $L = 8, 12, 16, 24, 32, 48, 64, 128, 256$ from top to bottom

We now take the simple average of $\epsilon_1(r,L)$ over r in the mentioned region and get values of $\epsilon_1(L)$. In the same manner as before, we could now take $\epsilon_1 = \lim_{L\to\infty}\epsilon_1(L) \approx 1.0473(1)$ which is consistent with the values obtained by other methods and coincides well with the exact value.

It should be noted that the accuracy of this method is not as good as that of the previous ones because we take simple numerical derivatives, which magnify all fluctuations. Nevertheless, the method of derivatives give us clear idea on the interval (r_{min}, r_{max}) of aspect ratio over which we can safely compute exponents. So, it is helpful as a supplement to the two methods described above.

6.3 Two and Three ISC's in an Open Strip

The same analysis of the data for probability of at least two ISC leads to the exponent $\epsilon_2 = -6.294(6)$ which is within two sigma from the exact value $-2\pi = -6.2832$. The corresponding exponent for the three ISC's are also computed $\epsilon_3 = -15.51(1)$. This value is well coincides with the exact value $-28/3\pi \approx -15.708$. We got relatively large deviation because the value of such probability decays very fast with the aspect ratio r and practically we take averages only over the rather small interval of aspect ratio $0.5 \leq r \leq 1.25$. The good agreement could be explained by the small corrections of order of $\exp(-2\pi r)$ to the leading behaviour [11].

7 Periodic Boundary Conditions

Periodic boundary conditions in vertical directions lead to different values of spanning probabilities. The geometry is equivalent to an open ended cylinder with the circle length $L_v = L$ and "horizontal" length $L_h = rL$. Clusters can span from the one cylinder end to another one also rotating around the cylinder. This leads to larger values of the one cluster spanning probability than in the case of open strip described in above. Actually, we obtain a spanning probability equal to 0.6365(1) which is larger than 0.5 in the case of the open strip and in a good comparison with the result of the paper [22] and more accurate. So, the spanning probability depends not only on the aspect ratio but on the boundary conditions as well [4,19,22]. As a consequence, the values of multiple ISC probabilities should be smaller than the corresponding ones for the open strip.

We applied both methods described in the previous section to the data analysis of the spanning probability $P(1, r; L)$ along an open ended cylinder.

Using the first method we compute the slope ϵ_1 of the logarithm of the spanning probability and get $\epsilon_1 = -0.6547(2)$. Fig. 6 shows the probability together with the best fit.

The second method based on the slope $\epsilon_1(L)$ calculation for each given lattice size $L = 8, 12, 16, 20, 24, 28, 32, 36, 40, 48, 64$ and then taking the limit of infinite lattice, finally leads to the value $\epsilon_1 = -0.65448(5)$. Both values are very close to the exact value $-5\pi/24 = -0.654498$. The second method produce much better accuracy.

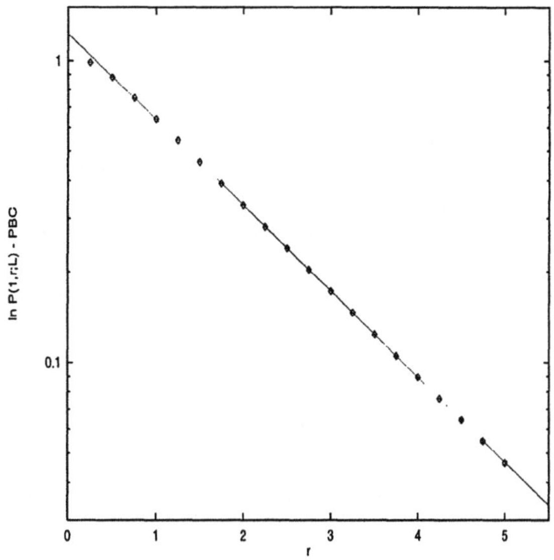

Fig. 6. Spanning probability $P(1,r)$ along cylinder as function of aspect ratio r denoted by open rhombs with error bars. Solid line is the best fit to probability in the interval $1.75 \leq r \leq 5.5$

Finite size corrections to the limiting values of the slope are similar to those for free boundaries. The only difference is in the finite size corrections of $\epsilon_1(L)$. The leading term, proportional to the inverse width vanishes, and the second power plays the main role, namely $\epsilon_1(L) = \epsilon_1 - 0.309(6)/L^2$. (We recall, that in the case of open boundaries the finite size corrections were $\epsilon_1(L) = \epsilon_1 + c/(L + L_0)$.)

The $1/L^2$ finite size corrections also hold for $\epsilon_2(L)$ and $\epsilon_3(L)$, whose values are $\epsilon_2(L) = -7.852(1)$ and $\epsilon_3(L) = -18.11(15)$, also quite close to the exact values $-5\pi/2 = -7.8540$ and $-35\pi/6 = -18.33$.

8 Incipient Spanning Clusters in Simple Cubic Lattice

Aizenman, in his seminal paper [6], also proposed the ISC exponents in dimension d from lower $d_l = 2$ to upper $d_u = 6$ critical dimensions

$$P_h(k,r) \propto \exp(-\alpha k^{\frac{d}{d-1}}) = \exp(-\epsilon_k r). \tag{13}$$

The preliminary analysis of the data obtained for the probabilities of clusters spanning between two opposite planes of the open cube (rL, L, L) at the critical site percolation threshold $p_c = 0.3116$ [34,35] are presented in the Table 4. We compute values of ϵ_k using the second method described above for system sizes $L = 8, 16, 32, 48, 64$ and aspect ratio up to $r = 7.25$, so some samples contain 1900544 sites. The averages were taken over 10^7 such samples.

A possible fit to the data in Table 4 is $\alpha_k \propto -\pi(k^{1.5} - 0.22k + 0.11)$, which supports Aizenman proposal. We cannot exclude, nevertheless, the power in exponent being less than $d/(d-1) = 1.5$.

In their preprint [36], Sen and Bhattacharjee proposed on the basis of their simulations that in the directed 3D percolation the ISC exponent depends with the same power 2 on the number of clusters k. Here we compute directly exponents as the slope of the logarithm of the corresponding probabilities, whereas in the Sen and Bhattacharjee preprint only values of probabilities for one fixed aspect ratio $r = 1$ were determined. We are not sure their analysis could distinguish the powers 1.5 and 2. We could only say that analysing our very precise data, we believe the value of power to be not larger than 1.5.

More detailed simulations and analysis are in progress.

Table 4. Exponents ϵ_k depending on the number k of ISC clusters in simple cubic site critical percolation

k	ϵ_k	σ
1	-2.7641	0.0004
2	-7.935	0.001
3	-14.68	0.01
4	-21.7	0.2

9 Discussion

We compute the values of the new exponents connected with the probabilities of a number of incipient spanning clusters in two dimensional percolation at criticality.

The results of comparison of the computed values with the exact ones are summarized in Table 5. Our data strongly support the exact results (6-8) for the number of incipient spanning clusters in two dimensional percolation developed by Cardy in [17,11] and, therefore, the validity of conformal invariance for multiple spanning probabilities.

Finite size corrections of the data computed with a high accuracy demonstrate that at criticality, corrections to all quantities analyzed depends on the inverse system size or on the inverse square size. We did not find any evidence for the irrelevant exponent $\theta_1 = 0.85$ as discussed in [22].

We stress the possibility of self duality for *finite* square lattices in the case of bond percolation.

We generalize finite size corrections developed by Ziff for the site percolation in a square geometry to any aspect ratio. We found that finite size

Table 5. Comparison of computed data with the exact results. Site percolation in two dimensional square lattice with free boundaries (FBC) and periodic boundaries (PBC)

k	Free boundaries		Periodic boundaries	
	α_k Exact	α_k Comp.	α_k Exact	α_k Comp.
1	$-\frac{\pi}{3} = -1.047197$	-1.0473(2)	$-\frac{5\pi}{24} = -0.654498$	-0.65448(5)
2	$-2\pi = -6.28318$	-6.294(6)	$-\frac{5\pi}{2} = -7.8540$	-7.852(1)
3	$-5\pi = -15.70796...$	-15.51(1)	$-\frac{35\pi}{6} = -18.33$	-18.11(15)

scaling of the probability of cluster spanning along a strip with open boundaries is described by a nonmonotonic function $b_0(r)$ which vanishes at the aspect ratio $r = 1/2$, or nearby.

Our preliminary data support Aizenman's proposal for the ISC exponent in three dimensional critical percolation. More data and analysis are necessary for definite check.

10 Acknowledgments

The author would like to thank D. Stauffer, A. Aharony, R. Ziff, J. Cardy, P. Butera and J.-P. Hovi for fruitful discussions. Special thanks to S. Kosyakov together with whom most of the results were computed. The author is thankful for the invitation to D.P. Landau and for kind hospitality to the Center of Simulational Physics, University of Georgia, where this paper was finished. This work was supported in part by grants from RFBR, INTAS, Centro Volta - Landau Network and NWO.

References

1. A. Bunde, S. Havlin, Eds., *Fractals and Disordered Systems*, (Springer, Berlin, 1996)
2. C.M. Fortuin, P.W. Kasteleyn, Physica **57** (1972) 536
3. D. Stauffer, A. Aharony. *Introduction to Percolation Theory*, 2nd ed. (Taylor & Francis, London, 1995)
4. R. Langlands, C. Pichet, P. Pouliot, Y. Saint-Aubin, J. Stat. Phys. **67** (1992) 553
5. R.M. Ziff, Phys. Rev. Lett. **69** (1992) 2670
6. M. Aizenman, Nucl. Phys. B [FS] **485** (1997) 551
7. G. Grimmett, *Percolation*, (Springer, New York, 1989)
8. C.-K. Hu and C.-Y. Lin, Phys. Rev. Lett., **77** (1996) 8
9. L.N. Shchur and S.S. Kosyakov, Int. J. Mod. Phys. C 8 (1997) 473, Nucl. Phys. B (Proc. Suppl.) **63A-C** (1998) 664
10. P. Sen, Int. J. Mod. Phys. C **7** (1996), Int. J. Mod. Phys. C **8** (1997) 229
11. J. Cardy, J. Phys. A **31** (1998) L105
12. S.S. Kosyakov, S.A. Krashakov, L.N. Shchur, Proc. Int. Conf. PDPTA'97, Las Vegas, Nevada, USA, Vol.2, 1170-1173 (Ed. H. R. Arabnia, 1997)

13. J. W. Essam, Rep. Prog. Phys. **43** (1980) 833
14. H. Kesten, *Percolation theory for mathematicians*, (Birkhäuser, Boston, 1982)
15. M. Aizenman and D.J. Barsky, Comm. Math. Phys. **108** (1987) 489
16. P.J. Reynolds, H.E. Stanley, and W. Klein, J.Phys. A **11** (1978) L199, Phys. Rev. B **21** (1980) 1223
17. J.L. Cardy, J. Phys. A **25** (1992) L201
18. J.R. Heringa, H.W.J. Blöte, A. Compagner, Int. J. Mod. Phys. C **3** (1992) 561
19. R. Langlands, P. Pouliot, Y. Saint-Aubin, Bulletin (New Series) of the American Mathematical Society, **30** (1994) 1
20. D. Stauffer, Physica A **242**, (1997) 1
21. R.M. Ziff, J. Phys. A **28** (1995) 1249 , J. Phys. A **29** (1995) 6479
22. J.-P. Hovi, A. Aharony, Phys. Rev. E **53** (1996) 235
23. M. Aizenman, in proceedings of STATPHYS 19, editor Hao Bailin (World Scientific, Singapore, 1996)
24. R.B. Griffiths and J.L. Lebowitz, J. Math. Phys. **9** (1968) 1284
25. O.A. Vasilyev and L.N. Shchur, unpublished
26. L.N. Shchur, H.W.J. Blöte and J.R. Heringa, Physica A **241** (1997) 579
27. L.N. Shchur and H. Blöte, Phys. Rev. E **55** (1997) R4905
28. R.M. Ziff, Computers in Physics **12** (1998) 385
29. L.N. Shchur and P. Butera, Int. J.Mod. Phys C **9** (1998) 607
30. J. Hoshen, R. Kopelman, Phys. Rev. B **14** (1976) 3438
31. L.N. Shchur and S.S. Kosyakov, unpublished
32. B.I. Shklovskii, A.L. Efros, *Electronic Properties of Doped Semiconductors. 5. Percolation Theory* (Springer-Verlag, Berlin, 1984) pp. 95-136
33. R.M. Ziff, Phys. Rev. E **54** (1996) 2547
34. P. Grassberger, J. Phys. A **25** (1992) 5867
35. N. Jan and D. Stauffer, Int. J. Mod. Phys. **C9** (1998) 341
36. P. Sen and S.M. Bhattacharjee, preprint cond-mat/9804282

Part III

Spin Glasses

A Bivariate Multicanonical Monte Carlo of the 3D $\pm J$ Spin Glass

N. Hatano and J. E. Gubernatis

Theoretical Division, Los Alamos National Laboratory
Los Alamos, NM 87545, USA

Abstract. A bivariate multicanonical Monte Carlo simulation of the three–dimensional $\pm J$ Ising spin glass is described. The autocorrelation time is approximately proportional to the system size, which is a great improvement over previous spin–glass simulations. The Binder plot indicates the critical temperature $T_c \simeq 1.3$. The order–parameter distribution function $P(q)$ exhibits a feature of the droplet picture of the low–temperature phase.

1 Introduction

It has been a few decades since spin glasses first attracted much attention [1]. Although there have been many extensive studies, theoretical, numerical, and experimental, there still remain a few unsettled fundamental problems. In the present paper, we report the results of our recent Monte Carlo simulation of the three–dimensional $\pm J$ Ising spin glass, focusing on two issues: estimation of the spin–glass transition temperature and the nature of the low–temperature phase [2].

First, in the next section, we give a short review of the current status of spin–glass studies. In Sect. 3, we describe our simulation method, namely a bivariate multicanonical Monte Carlo method. Monovariate multicanonical methods [3,4] have been applied to spin glasses before [5–7]; we found, however, that using a bivariate version is essential in reducing the correlation time of the simulation. We show that the autocorrelation time is approximately proportional to the system size. Since the output of multicanonical simulations is the density of states, we can access to any temperatures in principle. Thus we were able to explore low–temperature regions which have been hard to access by other Monte Carlo methods. We present our Monte Carlo results in Sect. 4. Our Binder plot clearly indicates the existence of a spin–glass transition at the critical temperature $T_c \simeq 1.3$. In addition, the order–parameter distribution function $P(q)$ exhibits a feature of the droplet picture of the low–temperature phase: $P(0)$ decreases at $T = 0.3$ as the system size is increased. The both results are not completely consistent with previous Monte Carlo results. We discuss possible reasons of the inconsistency in the last section.

Springer Proceedings in Physics, Volume 85
Computer Simulation Studies in Condensed-Matter Physics XII
Eds.: D.P. Landau, S.P. Lewis and H.-B. Schüttler

2 Spin Glass

2.1 The $\pm J$ Model and the Overlap Order Parameter

Spin glasses are realized typically in alloys of a magnetic metal and a non-magnetic metal [1]. Because the RKKY exchange interaction between the magnetic moments alternates in sign as a function of the distance, some of the magnetic–moment pairs interact ferromagnetically, some antiferromagnetically.

A simple theoretical model to which we focus our attention is the $\pm J$ Ising spin glass. Its Hamiltonian is given by [8]

$$\mathcal{H} \equiv -\sum_{\langle i,j\rangle} J_{ij}\sigma_i\sigma_j. \tag{1}$$

Here the Ising spins σ_i fluctuates thermodynamically, while each exchange interaction J_{ij} is quenched to $\pm J$ randomly. The summation runs over the nearest–neighbor pairs of a cubic lattice.

If all the exchange interactions were positive, there would be a ferromagnetic phase transition at a finite temperature, below which the system falls into one of the two global minima of the free energy, namely the ordered states. "Frustration" of the interactions drastically changes the situation [9]. The random interactions generate many local minima in the free energy landscape, each of which may represent a seemingly random configuration of the spins. However, the spins may be frozen into such a configuration at low temperatures; this spin freezing is the essence of the spin–glass "order."

A theoretical measure of the spin–glass phase is called the overlap order parameter [8]. To define it, we duplicate the random exchange interactions in (1) and change the Hamiltonian to

$$\mathcal{H} \equiv -\sum_{\alpha=1,2}\sum_{\langle i,j\rangle} J_{ij}\sigma_i^{(\alpha)}\sigma_j^{(\alpha)}, \tag{2}$$

where the superscript of the spin variables indicates each replica of the system. The spins of one replica and those of the other are thermodynamically independent. Nonetheless, we can expect that the spin configurations of the two replicas are similar to each other at low temperatures, if there is a thermodynamic state which the system is frozen into. Thus we define the order parameter as the Hamming distance of the spin configurations of the replicas [8]:

$$q = \frac{1}{N}\sum_{i=1}^{N}\sigma_i^{(1)}\sigma_i^{(2)}. \tag{3}$$

Here N is the number of the lattice sites. One may regard that the two replicas represent the same system at two moments in time with infinite interval [8];

hence the parameter q measures how strongly the spins are frozen into a configuration.

The thermodynamic average $\langle\cdots\rangle$ in the following is calculated with respect to the replica Hamiltonian (2). In addition, we calculate the random average $[\cdots]_{\rm av}$ over various samples $\{J_{ij}\}$ to obtain the final estimates of physical quantities.

2.2 Theoretical Issues

Because of the very nature of the spin–glass order, it is generally hard to equilibrate spin glasses in simulations and in experiments. After a few decades of research, there are still unresolved issues on equilibrium properties of spin glasses. Fundamental theoretical questions that we address here regarding spin glasses in equilibrium are: (i) what is the critical temperature $T_{\rm c}$, below which $[\langle q\rangle]_{\rm av} \neq 0$? (ii) what is the nature of the low–temperature phase?

It is now widely accepted that there is no spin–glass phase transition in two dimensions. It has been harder to determine the existence of the transition in three dimensions.

A well-used method of estimating the critical temperature of various systems, and particularly spin glasses, is the Binder parameter given by [10]

$$g_L \equiv \frac{3}{2}\left[1-\frac{\langle q^4\rangle}{3\langle q^2\rangle^2}\right]_{\rm av}, \tag{4}$$

where L denotes the linear system size. Because of the dimensionless combination of the second moment and the fourth moment, the Binder parameter is expected to be independent of the system size at fixed points, *i.e.* $T=0, T_{\rm c}, \infty$. The order–parameter distribution is a Gaussian at $T=\infty$; hence $\langle q^4\rangle = 3\langle q^2\rangle^2$ and $g_L=0$. If the order parameter can take only two values $\pm q_0$ at $T=0$ as in usual ferromagnets, we have $\langle q^4\rangle = \langle q^2\rangle^2 = q_0^4$ and hence the Binder parameter is $g_L=1$. At $T=T_{\rm c}$, the Binder parameters for different system sizes are expected to cross at a nontrivial value. In the high–temperature phase, $g_L(T)$ decreases towards zero as $L\to\infty$, while it increases towards unity in the low–temperature phase. Thus the crossing point of $g_L(T)$ should give the critical temperature.

Unfortunately, in earlier studies of spin glasses, the numerical results for the Binder parameter did not show the expected behavior; the Binder parameter $g_L(T)$ for different L appeared to collapse into one curve at low temperatures [1,11], which made one even doubt the applicability of the Binder plot to spin glasses. The issue was settled only when two independent Monte Carlo studies several years ago yielded the expected behavior of the Binder parameter [12,13]. (In other words, the earlier numerical results turned out to be inaccurate at low temperatures, probably owing to slow dynamics of the simulations.) The crossing point of the new studies was $T_{\rm c}\simeq 1.1$. We stress, however, that the Binder–parameter crossing in these studies was still subtle compared to our result. As we describe below, our results show a clearly visible crossing point near $T_c \simeq 1.3$. The crossing point is clearer even than in the two new studies.

The other theoretical issue is the nature of the low–temperature phase. This is the subject of a long-standing and unresolved controversy between the mean–field picture and the droplet picture. The mean–field picture maintains the existence of an infinite number of global minima of the free energy, which is called the replica symmetry breaking [14]. This is true for the mean–field (Sherrington–Kirkpatrick) model [15,16]; the question is whether this also applies to finite–dimensional models. The droplet picture, on the other hand, claims that only two global free–energy minima exist, although there must be many local minima [17,18]. According to this picture, the nature of the spin–glass transition is less exotic in three dimensions than in the mean–field model.

While many numerical simulations appear to support infinite degeneracy [19–21], a recent analytic treatment argues against it [22,23]. Note that the droplet picture actually is a theory concerning the *zero*–temperature fixed point; if the droplet picture is correct, it might be seen only near $T = 0$. Unfortunately, the slow dynamics of spin glasses has hindered numerical simulations from approaching the zero–temperature fixed point. A recent numerical study [24] demonstrated that results of the Migdal–Kadanoff approximation appear to indicate the mean–field picture near and below the critical point, but eventually come to support the droplet picture as $T \to 0$. Uncertainty, however, comes from the fact that the Migdal–Kadanoff approximation produces the droplet picture even for the mean–field model.

One way of distinguishing the two pictures is to measure the order–parameter distribution $P(q)$, which is given by

$$P(q) \equiv [\langle D(E,q) \rangle]_{\text{av}} \tag{5}$$

where $D(E,q)$ denotes the density of states with respect to the energy and the order parameter. If the droplet picture is correct, $P(q)$ should converge to two delta functions as $T \to 0$ and $L \to \infty$, because only two thermodynamic states are available at $T = 0$ according to this picture. If the mean–field picture, on the other hand, is correct, $P(q)$ should maintain a nontrivial form even as $T \to 0$. Hence many numerical simulations focus onto the size dependence of $P(0)$. The droplet picture predicts that $P(0)$ decreases as $L \to \infty$ at low enough temperatures, while the mean–field picture predicts that it converges to a finite value.

As was mentioned above, many simulation studies suggests the mean–field picture on the basis of their numerical estimates of $P(0)$ at as low as $T = 0.7$. Again, because of the slow dynamics of the simulations, it has been hard to lower the temperature further and to explore the vicinity of the zero–temperature fixed point. Using a multicanonical Monte Carlo method, we managed to reduce the difficulty of the slow dynamics significantly. Multicanonical simulations produce the density of states as output, from which we can calculate expectation values at any temperatures in principle. Our estimation of $P(0)$ at low temperatures suggests the droplet picture, as shown in the next section.

Let us add a few words regarding the correspondence between theories and experiments of spin glasses. As was stressed above, it generally takes very long time to equilibrate experimental samples. Hence many experiments nowadays feature nonequilibrium behavior of spin glasses. One might then say that the above theoretical issues regarding equilibrium properties, particularly the second one, are pedantic, if the equilibrium is not achievable in reality. However, the above issues are in fact linked to many different subjects of complex systems beyond spin glasses [14], and hence worth resolving. Furthermore, there are also efforts on the experimental side to quicken the equilibration process (*e.g.* with quantum fluctuation from transverse magnetic fields [25]) in the hope of achieving the true equilibrium of samples.

2.3 Simulation Methods

Attempts to resolve the above theoretical issues have been closely tied to the development of simulation methods. A simulation method that reduces the difficulty of the slow dynamics will be also useful in studying other complex systems.

One may classify the existing Monte Carlo methods for complex systems into two categories: cluster algorithms and extended–ensemble approaches. Swednsen and Wang [26] applied a cluster algorithm to spin glasses first in 1986 and saw successful results in two dimensions. However, cluster algorithms have been less successful in three dimensions.

In the last decade, many studies found extended ensembles useful in simulations of complex systems. Here the extended ensemble refers to an ensemble other than the canonical and micro-canonical ensembles. In spin glasses, canonical–ensemble simulations tend to be trapped in metastable states because of its strong emphasis on low–energy states at low temperatures. The main purpose of using extended ensembles is to provide paths out of metastable states via high–energy states without giving up sufficient statistics of low–energy states.

Among various extended–ensemble approaches is the multi-canonical Monte Carlo method [3,4]. In this method, the ensemble is defined so that any *macro*states appear at the same frequency in the course of the simulation. Berg and others [5–7] have applied this method to spin glasses, taking the weight of each spin configuration $\{\sigma\}$ either as

$$W(\{\sigma\}) \propto 1 \,/\, D\,(E(\{\sigma\})) \tag{6}$$

or

$$W(\{\sigma\}) \propto 1 \,/\, D\,(q(\{\sigma\})), \tag{7}$$

where $D(E)$ is the density of states with respect to the energy and $D(q)$ is the one with respect to the overlap order parameter. It has been argued [27] that "simulated tempering [28]," where replicas at different temperatures have an equal weight, is mathematically similar to taking the weight (6).

As we show in the following, we used for our simulation the bivariate weight

$$W \propto 1 \,/\, D(E, q). \tag{8}$$

Bivariate multicanonical Monte methods have been used in different contexts, *e.g.* protein simulations [29,30]. In preliminary runs, we found that the use of both E and q for spin–glass simulations is essential in improving the slow dynamics [2].

3 Multicanonical Monte Carlo Method

In this section, we describe the multicanonical Monte Carlo method in more detail. Let us first compare it with micro-canonical and canonical methods in the following way. Suppose that we carry out a simulation with the importance weight $W(E)$, keeping the histogram of the states with respect to E. After equilibration of the simulation, each microstate (spin configuration) is generated in the course of the simulation at a rate proportional to $W(E)$. The rate at which a *macro*state (thermodynamic state) with the energy E appears is proportional to $D(E)W(E)$, where $D(E)$ is the density of states.

In the micro-canonical simulation, we generate all microstates at the same rate. In other words, we set $W(E)$ = const., and hence the resulting histogram is proportional to the density of states: $h(E) \propto D(E)$ (see Fig. 1). In

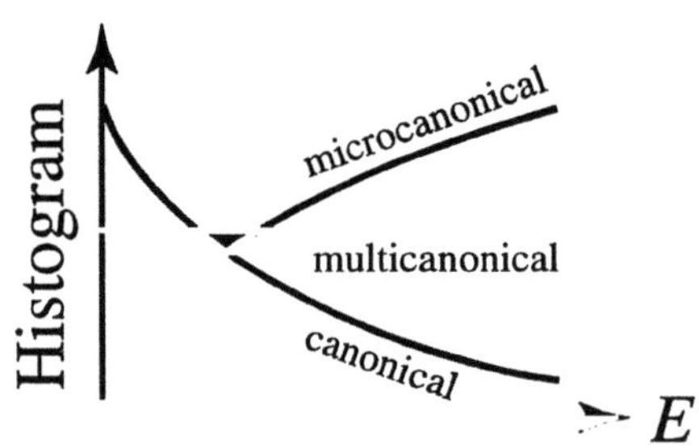

Fig. 1. A schematic of the histograms of states that would be generated in the micro-canonical, canonical, and multicanonical simulations

many cases, the density of states becomes very small in low–energy regions. Hence simulations based on the micro-canonical ensemble are generally not suitable for investigating low–temperature properties; low–energy states that dominate the thermodynamic average at low temperatures are not generated frequently.

Algorithms of importance sampling, or the canonical method, were developed to overcome this difficulty. In the canonical method, states are generated with the importance weight $W(E) \propto e^{-\beta E}$, and hence the energy histogram is $h(E) \propto e^{-\beta E} D(E)$. Thus, in ideal cases, low–energy states appear at great frequency at low temperatures. In spin–glass simulations, however, the simulation tends to fall into a local minima of the free–energy landscape and stay there, because the system would have to be excited to a higher–energy state to go to the global minima.

In the multicanonical simulations, the importance weight is set to $W(E) \propto 1/D(E)$, and hence the histogram should become flat. (Of course, we cannot set $W(E) \propto 1/D(E)$ from the beginning, since we do not know the density of states *a priori*. However, there is a procedure that makes the importance weight converge to the reciprocal of the density of states [31].) The aim is to ensure better statistics of low–energy states than in the micro-canonical method and, at the same time, generate more high–energy states than in the canonical method so that the system can escape local free–energy minima.

In the ideal situation, the multicanonical simulation generates a free random walk in the macroscopic phase space, since all the macrostates should appear with the same probability. We can deduce from this that the autocorrelation time of the thermodynamic variable (E in the above case) should be proportional to the volume (the number of the sites). In the simulation of the $\pm J$ model, the energy change after each spin flip is of the order of J. On the other hand, the upper and lower bounds of the energy are of the order of NJ, where $N = L^3$ is the number of the sites. Hence the number of spin flips necessary for the random walk to cover the entire energy phase space is of the order of N^2. Thus the measured energy should decorrelate after N^2 Monte Carlo steps, or N Monte Carlo sweeps.

This was not achieved, however, in the previous attempts of the multicanonical simulations of spin glasses [5–7]. Berg *et al* reported $\tau \propto N^{2.8}$ for the importance weight (6) [5] and $\tau \propto N^{2.42}$ for the importance weight (7) [7]. (Here τ is measured in the unit of Monte Carlo *sweep*.) This suggests that the slow dynamics in the low–energy region was not removed completely in these monovariate multicanonical simulations. (Note, however, that Berg *et al* did not use the conventional definition of the autocorrelation time in measurement of τ. This might be another source of the above power laws.)

In the present study, we carried out the simulation using the bivariate importance weight (8). As shown in Fig. 2 (a), we almost achieved the ideal situation $\tau \propto N$. The random walk in the two–dimensional phase space is exemplified in Fig. 2 (b). The drawback of the bivariate simulation, on the other hand, is the need for large memory in the computation. The number of bins in the energy phase space is of the order of L^3, and so is that in the q phase space. Hence an array of the size L^6 is necessary to store each of

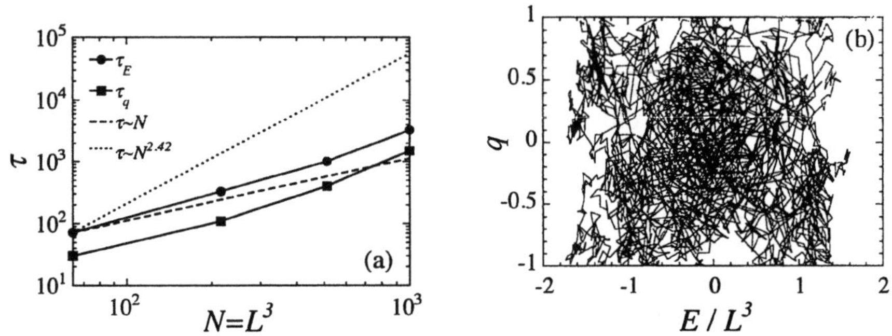

Fig. 2. (a) The system-size dependence of the autocorrelation time τ with respect to E and q. The dotted line represents the power law $\tau \propto N^{2.42}$, which was reported in [7]. (b) An example of the random walk in the phase space (E, q). The system size simulated here is $L = 4$. The time series of 2500 Monte Carlo sweeps after equilibration is plotted

the density of states $D(E, q)$, the histogram $h(E, q)$, and a work space. This amounts to about 30MB per sample for $L = 10$ and 400MB per sample for $L = 16$.

The actual algorithm of the multicanonical method has the following structure:

(A) Take a guess at the density of states, $D_0(E, q)$. Start the following loop with $i = 0$.

(B) The outer loop (multicanonical–iteration loop):

  (a) Set the importance weight as $W_i = 1/D_i$ Calculate the energy and the order parameter of the initial spin configuration, E_0 and q_0. Start the following loop with $j = 0$.

  (b) The inner loop (spin–update loop):

    (i) Choose a spin to be updated. Calculate the energy and the order parameter, E'_j and q'_j, assuming that the spin is actually flipped.

    (ii) Calculate the spin–flip probability as

$$P_{\text{flip}} = \min\left(1, \frac{W_i(E'_j, q'_j)}{W_i(E_j, q_j)}\right) \tag{9}$$

    (iii) Draw a random number $R \in [0, 1]$. If $R < P_{\text{flip}}$, flip the spin and make the substitution $E_{j+1} = E'_j$ and $q_{j+1} = q'_j$. Otherwise, keep the current spin configuration and make the substitution $E_{j+1} = E_j$ and $q_{j+1} = q_j$.

    (iv) Increment the histogram bin $h_i(E_{j+1}, q_{j+1})$ by one.

    (v) Increment the number of steps, j, by one and go to the step (i) until j exceeds a pre-determined number.

  (c) Measure a physical quantity as

$$\langle Q \rangle_i = \frac{1}{Z_i} \int dE \int dq\, Q(E, q) e^{-\beta E} h_i(E, q) D_i(E, q), \tag{10}$$

where the partition–function estimate is given by

$$Z_i = \int dE \int dq\, e^{-\beta E} h_i(E,q) D_i(E,q). \tag{11}$$

(d) Update the guess of the density of states as follows: [31]

$$D_{i+1}(E,q) = \frac{D_i(E,q)\,(h_i(E,q)+1)}{\text{Normalizaion}}. \tag{12}$$

The normalization constant is chosen so that the integration of the density of states becomes unity. (An extra count is added to the histogram in (12) in order to avoid zero division in setting the next importance weight $W_{i+1} = 1/D_{i+1}$ [31].)

(e) Increment the number of multicanonical iterations, i, by one and go to the step (a), until the histogram meets certain conditions of being flat.

The step (iv) should be skipped until the simulation reaches the "equilibrium" for the importance weight W_i. In our simulation, we devoted to equilibration Monte Carlo sweeps of a few times as long as the correlation time.

It is also best to skip the step (c) until the histogram h_i becomes practically flat. In principle, the estimater (10) should be applicable even before the histogram becomes flat. With sufficient statistics, the histogram h_i would be proportional to $W_i(E,q)D(E,q) = D(E,q)/D_i(E,q)$, where $D(E,q)$ is the true density of states. Thus the estimater (10) would give

$$\langle Q \rangle_i \simeq \frac{1}{Z_i} \int dE \int dq\, Q(E,q) e^{-\beta E} D(E,q), \tag{13}$$

with the partition function

$$Z = \int dE \int dq\, e^{-\beta E} D(E,q). \tag{14}$$

In practice, however, a limited number of Monte Carlo steps can cause large errors whereever h_i is small.

In the convergence process, the histogram is usually concentrated first in the mid part of the density of states and expands towards the upper and lower bounds of the energy until it becomes flat. Hence we determined that the histogram became flat, whenever the histogram count of the bin at the center of the density of states and that at the edges of the density of states are both close to the average histogram count.

As another remark, all physical quantities could be calculated after the whole simulation is over, if we stored the density of states itself on the storage disk. This would, however, require a huge disk space after simulating large number of random samples.

4 Numerical Results

We have carried out the bivariate multicanonical simulation of the systems of size $L = 4$ (838 samples), $L = 6$ (767 samples), $L = 8$ (719 samples) and

$L = 10$ (365 samples). The temperature dependence of the Binder parameter is shown in Fig. 3. The Binder–parameter crossing in this figure is much more prominent than in other numerical results. Although it is hard at this point to analyze the data including finite-size corrections, we can safely say that critical temperature is higher than 1.2.

The order–parameter distribution $P(q)$ at $T = 0.3$ is shown in Fig. 4. Although the statistical errors are still large, we can see the tendency of decreasing $P(0)$, which is the behavior predicted from the droplet picture. At $T = 0.7$, on the other hand, $P(0)$ does not show a definite tendency to decrease (Fig. 5), which may spuriously indicate the mean–field picture.

In the algorithm described in the previous section, we repeated the outer loop five times after we determined that the histogram became practically flat, so that we could detect erroneous estimates of the density of states. We also divide the inner loop into a few sections, from which we calculated the statistical errors. Note that errors of data at different temperatures of the same sample are generally correlated in multicanonical simulations.

5 Discussions

As we have seen, the results of our multicanonical Monte Carlo calculation are in disagreement with some of previous results. First, $P(q)$ near $q \simeq 0$ appears to be much smaller at low temperatures. Second, the Binder parameter appears to be greater in the low-temperature phase and this disagreement grows as the system size is increased.

These disagreements may be understood by the following scenario: If there are many local free-energy minima with small q, simulations may be trapped by them and thus overestimate $P(q)$ near $q = 0$, particularly when the system

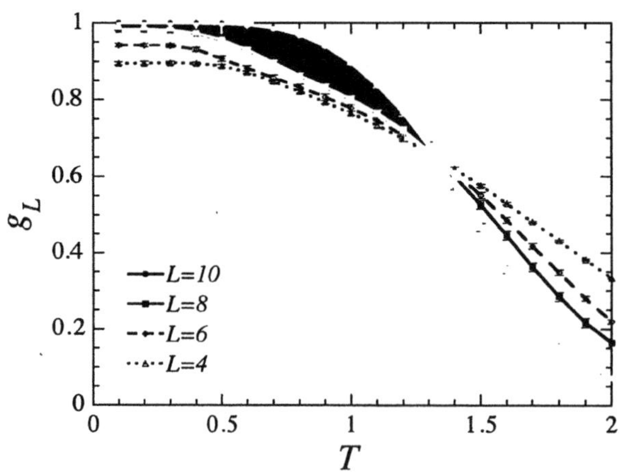

Fig. 3. The Binder plot for the 3D $\pm J$ Ising spin glass

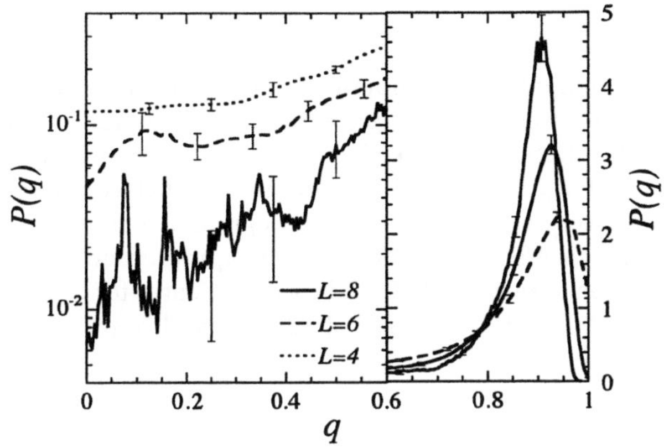

Fig. 4. The order–parameter distribution $P(q)$ at $T = 0.3$. The error bars are shown only for a part of the data points

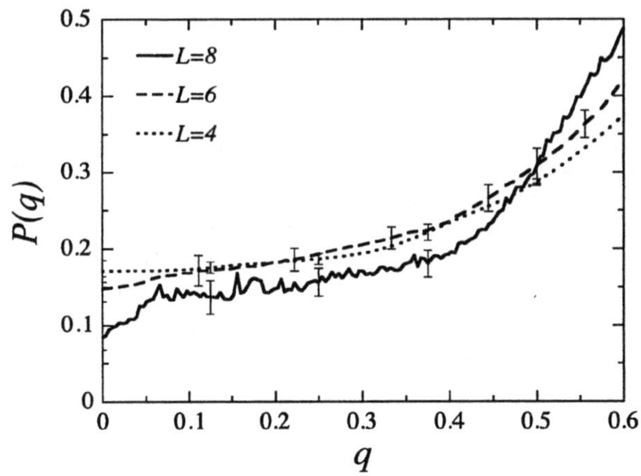

Fig. 5. The order–parameter distribution $P(q)$ at $T = 0.7$. The error bars are shown only for a part of the data points

size is increased. This would result in underestimating the Binder parameter of large systems at low temperatures. This scenario is supported by a recent study, where Sandvik[32] carried out a series of simulated annealing calculations with different annealing rates and estimated $P(q)$ in each annealing. He showed that the ground-state estimate of $P(0)$ is greater for faster annealing than for slower annealing (which must be more precise). His results suggest that simulations without substantial equilibration could overestimate $P(0)$.

Acknowledgments

We gratefully acknowledge computational support from the Applied Mathematics program of the Department of Energy for the use of the massively parallel computers at NERSC, from the Center for Nonlinear Studies (CNLS) at Los Alamos National Laboratory for the use of its parallel computer Avalon, and from the Laboratory Directed Research program at Los Alamos for some computer time on the ASCI Blue Mountain computer. The use of the random–number generator library SPRNG of National Center for Supercomputing Applications (NCSA) at University of Illinois is also gratefully acknowledged.

References

1. Binder, K., Young, A. P. (1986) Spin-Glasses: Experimental Facts, Theoretical Concepts, and Open Questions. Rev. Mod. Phys. **58**, 801–976
2. Preliminary results have been reported in the 8th Tohwa University International Symposium. See the proceedings, Hatano, N., Gubernatis, J. E., Bivariate Multicanonical Monte Carlo of the 3D $\pm J$ Spin Glass. In: Tokuyama, M. (Ed.) Slow Dynamics in Complex Systems. American Institute of Physics, Maryland, to be published
3. Berg, B., Neuhaus, T. (1991) Multicanonical Algorithms for First Order Phase Transitions. Phys. Lett. B **267**, 249–253
4. Berg, B., Neuhaus, T. (1992) Multicanonical Ensemble: A New Approach to Simulate First Order Phase Transitions. Phys. Rev. Lett. **68**, 9–12
5. Berg, B.A., Hansmann, U., Celik, T. (1994) Ground–State Properties of the Three–Dimensional Ising Spin Glass. Phys. Rev. B **50**, 16444–16452
6. Berg, B.A. (1996) Multicanonical Recursions. J. Stat. Phys. **82**, 323–342
7. Berg, B.A., Janke, W. (1998) Multioverlap Simulations of the 3D Edwards–Anderson Ising Spin Glass. Phys. Rev. Lett. **80**, 4771–4774
8. Edwards, S. F., Anderson, P. W. (1975) Theory of Spin Glasses. J. Phys. F: Metal Phys. **5**, 965–974
9. Toulouse, G. (1977) Theory of the frustration effect in spin glasses: I. Comm. Phys. **2**, 115–119
10. Binder, K. (1981) Finite Size Scaling Analysis of Ising Model Block Distribution–Functions. Z. Phys. B **43**, 119–140
11. For example, Bhatt, R. N., Young, A. P., Numerical Studies of Ising Spin Glasses in 2, 3, and 4 Dimensions. Phys. Rev. B **37**, 5606–5614
12. Kawashima, N., Young, A. P. (1996) Phase Transition in the Three-Dimensional $\pm J$ Ising Spin Glass. Phys. Rev. B **53**, R484–R487
13. Hukushima, K., Takayama, H., Nemoto, K. (1996) Application of an Extended Ensemble Method to Spin Glasses. Int. J. Mod. Phys. C **7**, 337–344
14. For a review, Mézard, M., Parisi, G., Virasoro, M. (1987) Spin Glass Theory and Beyond. World Scientific, Singapore
15. Sherrington, D., Kirkpatrick, S. (1975) Solvable Model of a Spin–Glass. Phys. Rev. Lett. **35**, 1792–1796
16. Parisi, G. (1979) Infinite Number of Order Parameters for Spin–Glasses. Phys. Rev. Lett. **43**, 1754–1756

17. Fisher, D. S., Huse, D. A. (1986) Ordered phase of Short–Range Ising Spin–Glasses. Phys. Rev. Lett. **56**, 1601–1604
18. Fisher, D. S., Huse, D. A. (1988) Equilibrium behavior of the Spin–Glass Ordered Phase. Phys. Rev. B **38**, 386–411
19. Marinari, E., Parisi, G., Ruiz-Lorenzo, J., Ritort, F. (1996) Numerical Evidence for Spontaneously Broken Replica Symmetry in 3D Spin Glasses. Phys. Rev. Lett. **76**, 843–846
20. Iñiguez, D., Marinari, E., Parisi, G., Ruiz-Lorenzo, J. (1997) 3D Spin Glass and 2D Ferromagnetic XY Model: A Comparison. J. Phys. A: Math. Gen. **30**, 7337–7347
21. Marinari, E. (1999) Numerical Evidence for Continuity of Mean Field and Finite Dimensional Spin Glasses. Phys. Rev. Lett. **82**, 434–437
22. Newman, C. M., Stein, D. L. (1996) Non-Mean–Field Behavior of Realistic Spin Glasses. Phys. Rev. Lett. **76**, 515–518
23. Newman, C. M., Stein, D. L. (1998) Simplicity of State and Overlap Structure in Finite-Volume Realistic Spin Glasses. Phys. Rev. E **57**, 1356–1366
24. Moore, M. A., Bokil, H., Drossel, B. (1998) Evidence for the Droplet Picture of Spin Glasses. Phys. Rev. Lett. **81**, 4252–4255
25. Wu, W., Ellman, B., Rosenbaum, T. F., Aeppli, G., Reich, D. H. (1991) From Classical to Quantum Glass. Phys. Rev. Lett. **67**, 2076–2079
26. Swednsen, R. H., Wang, J.-S. (1986) Replica Monte Carlo Simulation of Spin–Glasses. Phys. Rev. Lett. **57**, 2607–2609
27. Hansmann, U., Okamoto, Y. (1996) Monte–Carlo Simulations in Generalized Ensemble: Multicanonical Algorithm versus Simulated Tempering. Phys. Rev. E **54**, 5863–5865
28. Marinari, E., Parisi, G. (1992) Simulated Tempering: A New Monte Carlo Scheme. Europhys. Lett. **19**, 451–458
29. Higo, J., Nakajima, N., Shirai, H., Kidera, A., Nakamura, H. (1997) Two–component Multicanonical Monte Carlo Method for Effective Conformation Sampling. J. Comp. Chem. **16**, 2086–2092
30. Iba, Y., Chikenji, G., Kikuchi, M. (1998) Simulation of Lattice Problems with Multi-Self-Overlap Ensemble. J. Phys. Soc. Jpn. **67**, 3327–3330
31. Smith, G.R., and Bruce, A.D. (1995) A Study of the Multi-Canonical Monte Carlo Method. J. Phys. A: Math. Gen. **28**, 6623–6643
32. Sandvik, A. (1999) Comment on "Evidence for Nontrivial Ground–State Structure of 3d $\pm J$ Spin Glasses." Europhys. Lett., to be published

Monte Carlo Simulation of the Three-Dimensional Ising Spin Glass

M. Palassini[1] and S. Caracciolo[2]

[1] Department of Physics, University of California, Santa Cruz, California 95064
[2] Scuola Normale Superiore and INFN, 56100 Pisa, Italy

Abstract. We study the 3D Edwards–Anderson model with binary interactions by Monte Carlo simulations. Direct evidence of finite–size scaling is provided, and the universal finite–size scaling functions are determined. Using an iterative extrapolation procedure, Monte Carlo data are extrapolated to infinite volume up to correlation length $\xi \approx 140$. The infinite volume data are consistent with both a continuous phase transition at finite temperature and an essential singularity at finite temperature. An essential singularity at zero temperature is excluded.

1 Introduction

Understanding the thermodynamics of spin glasses represents a challenging unsolved problem for statistical and computational physics [1]. The very existence of a phase transition in the 3D Ising spin glass is still unclear. Previous Monte Carlo (MC) simulations [2–6] give a certain evidence of a $T_c \neq 0$ continuous transition with power-law divergence of ξ at $T \to T_c^+$. However, they cannot exclude completely neither an exponential divergence at $T \to 0^+$ [2–4], nor a line of critical points at $T \leq T_c \neq 0$ [4,6,7], which implies an exponential divergence at $T \to T_c^+$. Furthermore, the important topic of finite-size scaling (FSS) has not been much investigated for spin glasses.

In this paper, we study the $3D$ Ising spin glass by MC simulations of moderate–size systems in the paramagnetic phase (less hampered by long equilibration times), using the powerful FSS method of [8]. In Sect. 2 details on the simulations are given. In Sect. 3 a direct evidence of FSS, independent of the nature of the divergence of ξ, is provided, and the *universal* FSS functions are determined for the first time [9]. The MC data are extrapolated to infinite volume up to ξ one order of magnitude larger than before. In Sect. 4 the critical behavior of the extrapolated data is analyzed.

2 Model and Simulation

We consider the $3D$ Edwards–Anderson model, whose Hamiltonian is

$$\mathcal{H} = -\sum_{\langle xy \rangle} \sigma_x J_{xy} \sigma_y \tag{1}$$

where σ_x are Ising spins on a simple cubic lattice of linear size L with periodic

Springer Proceedings in Physics, Volume 85
Computer Simulation Studies in Condensed-Matter Physics XII
Eds.: D.P. Landau, S.P. Lewis and H.-B. Schüttler

boundaries, and J_{xy} are independent random interactions taking the values ± 1 with equal probability. The sum runs over pairs of nearest neighbor sites.

We simulate with the heath–bath algorithm many samples of the model 1 with different J_{xy}. From two independent replicas (σ, τ) with the same J_{xy}, we measure the overlap $q_x = \sigma_x \tau_x$ and $q = L^{-3} \sum_x q_x$. We compute the *second-moment* correlation length using the following definition:

$$\xi(T,L) = \left(\frac{S(0)/S(p) - 1}{4 \sin^2(p/2)} \right)^{1/2} \tag{2}$$

where $S(k)$ is the Fourier transform of the overlap correlation function

$$S(k) = \sum_r e^{ik \cdot r} \langle q_x q_{x+r} \rangle , \tag{3}$$

(the arguments T, L are omitted) and $p = (2\pi/L, 0, 0)$ is the smallest non–zero wave vector. We also measure the spin–glass susceptibility $\chi_{SG}(T,L) \equiv L^3 \langle q^2 \rangle = S(0)$. The runs are done on a Cray T3E with a fast code (see [9] for details) that exploits the high parallelism of spin glass simulations, combining multi–processor parallelism with an efficient multi–spin coding technique. Average speed on a single processor (DEC Alpha EV5, 600 MHz) is 4.5×10^7 spin updates per second, and we typically used 32 to 128 PEs. We simulated 104 pairs (T, L), $(T, 2L)$ with L between 4 and 48.

3 Finite Size Scaling Analysis

According to the FSS hypothesis [10], if $\mathcal{O}(T,L)$ is some long-distance observable (for example, $\xi(T,L)$ or $\chi_{SG}(T,L)$), then

$$\frac{\mathcal{O}(T,L)}{\mathcal{O}(T,\infty)} = f_{\mathcal{O}}\Big(\xi(T,\infty)/L\Big) \tag{4}$$

where $f_{\mathcal{O}}$ is a universal function and corrections to FSS are neglected. From (1) we obtain a useful relation involving only finite–volume observables

$$\frac{\mathcal{O}(T,2L)}{\mathcal{O}(T,L)} = F_{\mathcal{O}}\Big(\xi(T,L)/L\Big) \tag{5}$$

where $F_{\mathcal{O}}$ is another universal function. As shown in Fig.1a,b and Fig.2, our data for the observables χ_{SG}, ξ, $q_4 \equiv \langle q^4 \rangle$ and $F \equiv V^{-3} S(p)$ verify the ansatz (2). This provides a direct test of the FSS hypothesis. Small systematic deviations, due to corrections to FSS, appear in Fig.2b for $L = 5$ and $\xi(T,L)/L \equiv x > 0.4$. We emphasize that FSS is not assumed *a priori* and that no adjustable parameters are contained in (2). Furthermore, no particular dependence of the observables on the temperature is assumed.

We fit the data in Fig.1a,b to two suitable functions $F_{\chi_{SG}}$, F_ξ and then extrapolate the pair (ξ, χ_{SG}) from $L \to 2L \to 2^2 L \to \ldots \to \infty$ using (2)

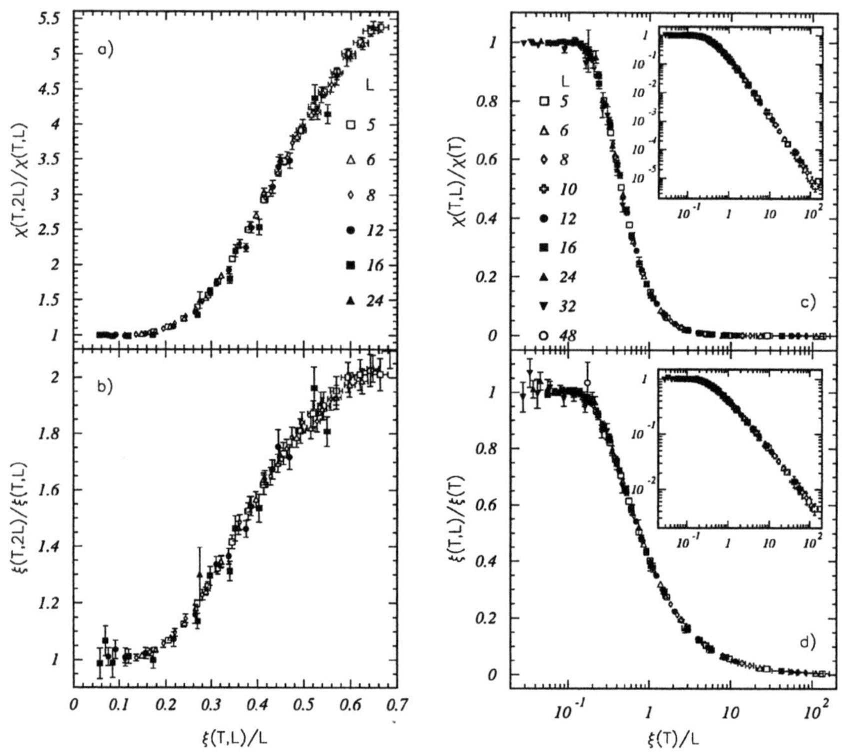

Fig. 1. **(a,b)** FSS plot with the form in (2) for **(a)** $\mathcal{O} = \chi_{SG}$ and **(b)** $\mathcal{O} = \xi$. Error bars (estimated with a jackknife procedure) are one standard deviation. **(c,d)** FSS plot with the form in (1) for **(c)** $\mathcal{O} = \chi_{SG}$ and **(d)** $\mathcal{O} = \xi$. See [8] for how to estimate error bars of the extrapolated data

iteratively. For any temperature, we verify that extrapolations from different L agree within the error bars. In this procedure, we assumed implicitly that (2) with a given function $F_{\mathcal{O}}$ will continue to hold as $L \to \infty$. This assumption could fail if there is a crossover at large L. However, at high T extrapolations from small L are consistent with data from large L, which have almost no finite-size effects, and thus a crossover is unlikely. We have a good control on the extrapolated data up to $\xi \approx 140$; at lower temperatures the statistical errors become quite large, and the data are more sensitive to FSS corrections.

In Fig.1c,d we show that with our extrapolated data (1) is satisfied remarkably well, providing a further test of the method. If $\mathcal{O} \sim \xi^{\gamma_{\mathcal{O}}/\nu}$ as $\xi \to \infty$, then $f_{\mathcal{O}}(x)$ in (1) must satisfy $f_{\mathcal{O}}(x) \sim x^{-\gamma_{\mathcal{O}}/\nu}$ as $x \to \infty$. As shown in the insets of Fig.1 (c,d), our curves indeed have a power-law asymptotic decay. We emphasize that all the scaling functions in Figs.1 and 2 are *universal.*

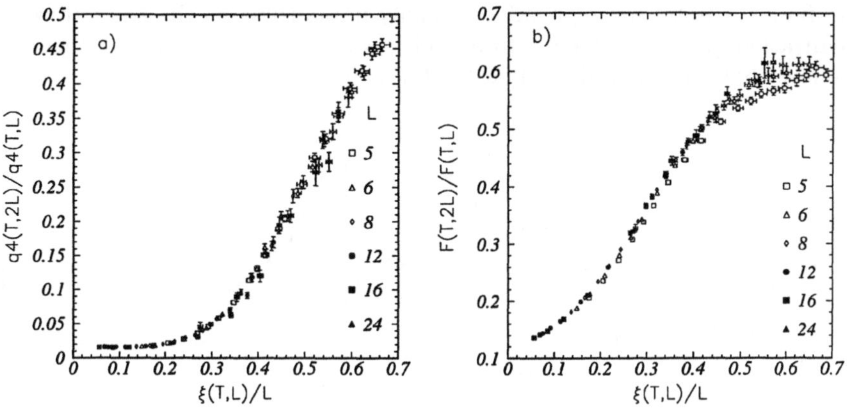

Fig. 2. FSS plot with the form in (2) for **(a)** $\mathcal{O} = q_4 \equiv \langle q^4 \rangle$ and **(b)** $\mathcal{O} = F \equiv V^{-3}S(p)$, $|p| = 2\pi/L$

4 Nature of the Phase Transition

We now compare our extrapolated data with the three different scenarios compatible with previous simulations.

(i) *Power–law singularity at* $T_c \neq 0$. We fit our data to

$$\xi(T) = c_\xi\,(T - T_c)^{-\nu}\,\left[1 + a_\xi\,(T - T_c)^\theta\right] \tag{6}$$

$$\chi_{SG}(\xi) = b\,\xi^{2-\eta}\,\left[1 + d\,\xi^{-\Delta}\right] \tag{7}$$

with fixed θ and Δ. Without corrections to scaling ($a_\xi = d = 0$), the fit parameters show small *systematic* variations when we vary the fit interval. The fits stabilize with $1 \le \theta \le 2$ and $1 \le \Delta \le 1.5$, the preferred values being $\theta = 1.4$ and $\Delta = 1.3$ (*goodness of fit* parameter $Q > 0.6$ and $Q > 0.98$ resp.). Our estimates for the critical parameters are $T_c = 1.156 \pm 0.015$, $\nu = 1.8 \pm 0.2$, and $\eta = -0.26 \pm 0.04$. Corrections to scaling are important for $\xi \le 10$ (Fig.3).

(ii) *Essential singularity at* $T_c \neq 0$. Our data fit very well also to

$$\xi(T) = f_\xi\,\exp\!\left(g_\xi/(T - T_c)^\sigma\right)\,. \tag{8}$$

For $\xi \ge 3.8$ the best fit (shown in Fig.3b) gives $\sigma = 0.5 \pm 0.3$, $T_c = 1.08 \pm 0.04$ ($Q = 0.69$). Deviations from this fit for $\xi < 3$ are consistent with corrections to scaling of $\approx 10\%$. Since for an exponential singularity we expect multiplicative logarithmic corrections, we tried also the fit

$$\chi_{SG}(\xi) = b_l\,\xi^{2-\eta_l}\,(\log \xi)^r \tag{9}$$

obtaining $\eta_l = -0.36 \pm 0.03$ and $r = -0.36 \pm 0.06$ ($Q > 0.9$) (see Fig.3a).

(iii) *Essential singularity at* $T = 0$. Fitting our data to

$$\xi(T) = f_\xi\,\exp\!\left(g_\xi/T^\sigma\right) \tag{10}$$

we find $\sigma \geq 9$. Such a high value is highly implausible on the basis of renormalization group arguments, from which we expect $\sigma \approx 2$ [9]. We therefore believe that an essential singularity at $T = 0$ is excluded.

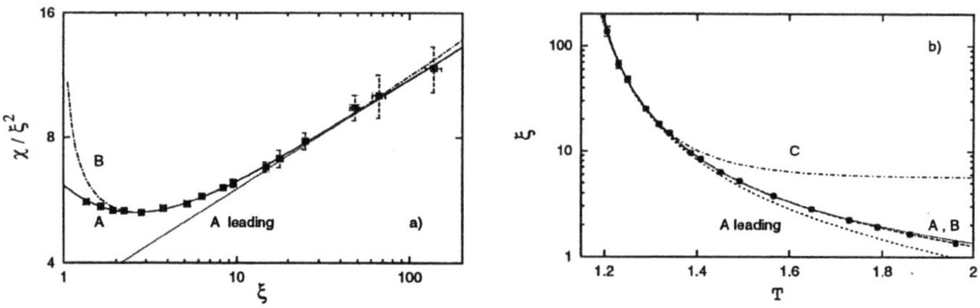

Fig. 3. Critical behavior of the infinite volume data. (**a**) Best fit to (7) for $\xi \geq 1.8$ (line A), leading term from the same fit (A leading), best fit to (9) for $\xi \geq 2.2$ (B). (**b**) Best fit to (6) for $\xi \geq 1.9$ (A), leading term from the same fit (A leading), best fit to (8) for $\xi \geq 3.8$ (B), and best fit to (10) for $\xi \geq 14$ (C)

Acknowledgements

We thank A. Pelissetto and A.P. Young for useful discussions. This work was supported by the INFM Parallel Computing Initiative.

References

1. K. Binder and A.P. Young, Rev. Mod. Phys. **58**, 801 (1986); E. Marinari, G. Parisi and J. Ruiz-Lorenzo, in *Spin Glasses and Random Fields*, edited by A. P. Young (World Scientific, Singapore, 1997).
2. A.T. Ogielski and I. Morgenstern, Phys. Rev. Lett. **54**, 928 (1985); A.T. Ogielski, Phys. Rev. B **32**, 7384 (1985).
3. E.Marinari, G.Parisi and F.Ritort, J. Phys. A **27** (1994).
4. N. Kawashima and A.P. Young, Phys. Rev. B **53**, R484 (1996).
5. E. Marinari, G. Parisi and J.J. Ruiz-Lorenzo, Phys. Rev. B **58**, 14852 (1998);
6. B.A. Berg and W. Janke, Phys. Rev. Lett. **80**, 4771 (1998); W. Janke, B.A. Berg and A. Billoire, cond-mat/9811423.
7. D. Iniguez, E. Marinari, G. Parisi and J.J. Ruiz-Lorenzo, J. Phys. A **30** 7337 (1997).
8. S. Caracciolo, R.G. Edwards, S.J. Ferreira, A. Pelissetto and A.D. Sokal, Phys. Rev. Lett. **74**, 2969 (1995).
9. M. Palassini and S. Caracciolo, submitted
10. M.N. Barber, in *Phase Transitions and Critical Phenomena*, vol. 8, edited by C. Domb and J.L. Lebowitz (Academic Press, London, 1983).

Part IV

Computer Simulation Methods

Molecular Dynamics Simulations with IMD

J. Roth, J. Stadler, M. Brunelli, F. Gähler, J. Hahn, M. Hohl, C. Horn, R. Mikulla, G. Schaaf, and H.-R. Trebin

Institut für Theoretische und Angewandte Physik, Universität Stuttgart, Pfaffenwaldring 57, 70550 Stuttgart, Germany

Abstract. We describe IMD (**ITAP Molecular Dynamics**), a software package for classical molecular dynamics simulations on massively-parallel computers. IMD is a general purpose program for short-ranged interactions. It contains special features for the investigation of the mechanical properties of solids. A representative application of IMD to shock waves in binary icosahedral quasicrystals is also given. The simulation has been carried out to demonstrate the feasibility of metacomputing (clustering of massively-parallel supercomputers over large distances).

1 Introduction

IMD is a software package designed for classical molecular dynamics (MD) simulations in two or three dimensions. The basic implementation of IMD and a description of release 1.0 has been published in a previous paper by Jörg Stadler et al.[1]. With the first release it was only possible to simulate particles interacting via pair potentials. Several thermodynamical ensembles were allowed, but the volume had to be kept constant.

Meanwhile IMD has been extended and improved. The current release 1.2[2] allows the particles to interact via pair, three-body and embedded atom (EAM[3]) potentials. The particles can be of an arbitrary number of types with different mass. The interaction potentials are read from tables and are therefore not limited to analytical expressions. The simulations can be carried out in microcanonical (NVE), and canonical (NVT, NPT, and NσT[1]) ensembles. The Nosé-Hoover thermostat is used for the canonical simulations. In addition to MD simulations it is possible to optimize the potential energy of a sample with the microconvergence method[5]. Mechanical boundary conditions allow the sample to be pulled apart or sheared at a constant rate or at certain time intervals.

IMD contains routines to compute self-correlation functions and to generate several crystal or quasicrystal structures. In addition there are special setups for fracture simulations to damp out shock waves emitted from a crack, and in the shock wave simulations to initiate a shock wave.

IMDs target platforms are RISC workstations and massively-parallel computers. It does not perform well on vector machines. IMD is written in C and uses MPI library for message passing. It has been adapted to metacomputing

[1] σ denotes uniaxial stress.

Springer Proceedings in Physics, Volume 85
Computer Simulation Studies in Condensed-Matter Physics XII
Eds.: D.P. Landau, S.P. Lewis and H.-B. Schüttler

via the PACX[6] library. For online visualization the VolRend[7] package has been extended to VolIMD which is permits communication with IMD.

This paper is organized as follows: In section 2 we will describe online visualization and the generation of pictures with IMD. In section 3 we will talk about metacomputing and in the final section present an example, the simulation of shock waves in quasicrystals

2 Online Visualization and the Generation of Pictures

Usually a MD simulation is carried out, data are collected during the run and stored for analysis *after* the simulation has been finished. With the large massively-parallel supercomputers such a scheme may no longer feasible since in many cases the amount of data is too large to be stored. A solution of this problem is to visualize the data online. More advanced schemes even permit to steer the simulation online but this is currently not supported by IMD although the basic interfaces are implemented already.

For online visualization the simulation is started interactively or in batch as usual. At any time during the simulation it is possible to connect a workstation to the supercomputer running the simulation and to get pictures or histograms. This is simply done by starting VolIMD on your workstation. The communication scheme implemented in IMD works in the following way: The central part of a MD simulation is the main loop for integration of the equations of motions. At a given time interval a function, `check_sockets`, is carried out which connects to a specified TCP/IP port at the workstation and looks whether certain flags have been set. If this is the case, the simulation program reacts as required and sends a picture or histogram of the kinetic or potential energy distribution to the workstation.

The output produced by the online visualization routines are raw data written in binary format. This has the advantage of being faster and more flexible than writing formatted output. Tools are available to convert the binary data into ppm[2]-files. In two dimensions ppm-files can be written directly, but the possibility to alter the pictures afterwards is lost.

IMD supports the generation of histograms of the kinetic and potential energy distribution in the sample. It has turned out that the kinetic energy is useful for example to visualize crack tips or shock waves, whereas the potential energy is sensitive to interface atoms or stacking faults. Since quasicrystals are rather complicated structures where the atoms have quite different potential energies it is also possible to store reference potential energies at the beginning of the simulation and to subtract them later.

[2] ppm: portable pixel map.

3 Metacomputing

Metacomputing is the extension of massively-parallel supercomputing to heterogeneous supercomputer clusters. Several applications are conceivable: if an application is too large for a single supercomputer and the computation center possesses several machines or if an application consists of a part more suitable for massively-parallel systems and another part more suitable for vector machines.

IMD is prepared for metacomputing with the PACX[3] library, developed at the HLRS in Stuttgart[6]. If necessary, PACX simply replaces standard MPI routines by new routines. Thus only minimal changes should be required for an existing massively-parallel program since some MPI routines are not yet implemented in PACX.

On each of the participating supercomputer nodes two processors (PE) are set aside for communication, one for sending and one for receiving. Figure 1 shows the communication between two hosts: The communication within one node is unchanged. But if a PE on one node wants to communicate with a PE on the other node, a two-step process is initiated: first the message is broadcasted locally to the communication PE, then it is sent to the communication PE on the other node and transmitted again over the local network to the destination PE. The communication with PACX was originally limited to two participants, meanwhile it has been extended to any number of supercomputer nodes. The communication PEs now talk to all other nodes, therefore two communication PE per node are still sufficient.

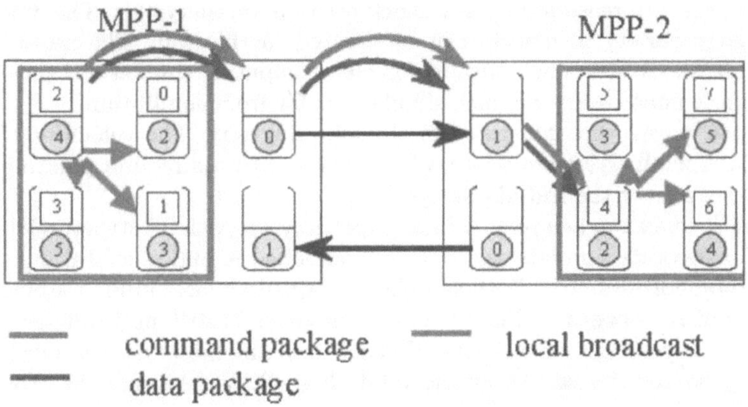

Fig. 1. Communication between two supercomputers. The labels in the squares and circles are the global and local processor addresses, respectively. The straight arrows indicate how the communication processors are connected

[3] PACX: **PA**rallel **C**omputer e**X**tension.

3.1 Special Requirements for Metacomputing

IMD has been written for massively-parallel supercomputers where the communication between the PE is very fast and and the communication time is short compared to the time required for a time step. For metacomputing this is no longer the case. The latency time for a message between two partitions on the same T3E is 3.5 ms, between Stuttgart and Manchester it is 45 ms, and between Stuttgart and Pittsburgh it is 75 ms! To avoid wrong results the computations has to wait until the inter-supercomputer communication has been finished. This is very unsatisfactory and a way to interlace calculation and communication has to be found.

The solution for a simulation program with domain decomposition (see Ref. [1]) is rather straightforward: The sample is divided into cells and distributed onto the PEs. Only the content of the surface cells of each PE has to be sent to the neighbor PEs. The loop over all cells on a PE has to be broken up into a central part not necessary to be communicated and into the surface part. During the exchange of the surface part a PE can start with the computations in the central part and then continue successively with the surface part that has been updated already.

4 Shock Waves in Quasicrystals and Crystals

There have been two reasons to study shock waves in quasicrystals: The first is that a large variety of defects can be created in this way. The second reason is that shock waves are suitable for metacomputing simulations since equilibration is unnecessary for such simulations far from equilibrium.

The shock waves were generated in the following way: The velocities of all atoms in a small slab were set to a large value with a direction pointing along the long axis of the rod-like sample.

We have studied two samples: A binary quasicrystal with TI structure[8], consisting of decorated prolate and oblate rhombohedra, and a crystal built of prolate rhombohedra only. Both samples contained about 1 million atoms and are as similar as possible. Their behavior, however, is strikingly different: First of all, the propagation velocity of shock front in the crystal is much larger than the velocity in the quasicrystal (Fig. 2). We assume that the aperiodicity of the quasicrystal decelerates the shock wave.

A second difference is observed for the strength of the material: the crystal is much stronger than the quasicrystal despite the similarity of the structures. Below a shock wave energy $I \approx 3$ in Lennard-Jones units no defects occur in the quasicrystal. Between $I \approx 3$ and $I \approx 13$ shear bands consisting of parallel lines of dislocations show up. Above $I \approx 13$ the sample becomes amorphous. The crystal is still perfect at a shock wave energy of $I = 12$, at $I = 18$ domains of disordered material are created, and at $I > 24$ it becomes amorphous. Shear bands have not been found in the crystal.

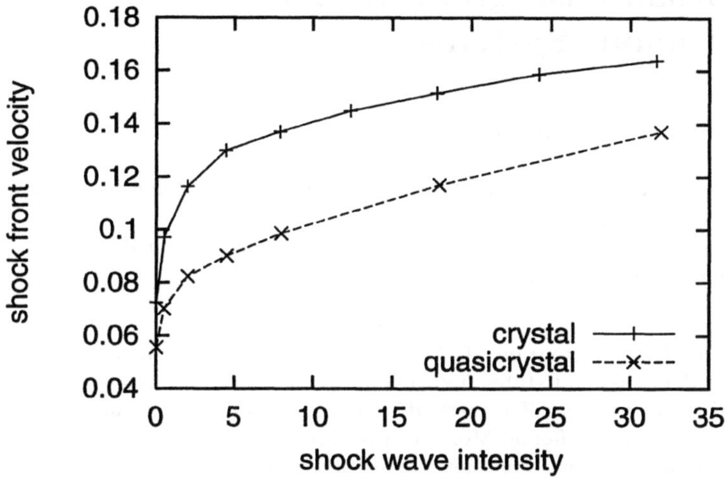

Fig. 2. Velocity of the shock wave (in arbitrary units) vs. strength of the shock wave, expressed in units of the average kinetic energy

To understand the differences between the strength of the crystalline and the quasicrystalline structure further simulations are necessary.

Acknowledgments

The help of the PACX group at the HLRS Stuttgart, led by Michael Resch and Edgar Gabriel, is greatfully acknowledged.

References

1. Stadler J., Mikulla R., Trebin H.-R. (1997) IMD: A Software Package for Molecular Dynamics Studies on Parallel Computers. Int. J Mod Phys C 8:1131
2. www.itap.physik.uni-stuttgart.de/~johannes/imd-home.html
3. Daw M.S., Baskes M.T. (1984) Embedded Atom Method: Derivation and Application to Impurities, Surfaces and other Defects in Metals. Phys Rev B 29:6443
4. Allen M.P., Tildesley D.J. (1987) Computer Simulation of Liquids, Oxford Science Publications, Oxford
5. Beeler J.R. (1983) Radiation Effects Computer Simulations, North Holland, Amsterdam, p 27
6. www.hlrs.de/structure/organisation/par/projects/pacx-mpi
7. www.uni-stuttgart.de/RUSuser/vis/People/roland/volrend/volrend.html
8. Roth J., Schilling R., Trebin H.-R. (1990) Stability of Monatomic and Diatomic Quasicrystals and the Influence of Noise. Phys Rev B 41:2735

Further Applications of the Broad Histogram Method for Continuous Systems

J. D. Muñoz[1,2] and H. J. Herrmann[1]

[1] Institute for Computer Applications 1, Stuttgart University, Pfaffenwaldring 27, D-70569 Stuttgart, Germany
[2] Permanent Address: Departamento de Física, Universidad Nacional de Colombia, Bogotá D.C., Colombia

Abstract. The Broad Histogram Method (BHMC) is a novel analysis technique for estimating the density of states from microcanonical averages related to some protocol of virtual movements. The method is completely general and can be applied in conjunction with many different Monte Carlo methods, including microcanonical, canonical and multicanonical simulations, to accelerate the estimation of the density of states, the specific heat, and similar thermodynamic quantities. In previous works [1,2] we proposed a way of extending the BHMC to systems with continuous degrees of freedom and applied these ideas to the study of the 3D XY-model. Here we summarize these ideas, present results for the 3D classical Heisenberg model, and show how to include an external field. Combined with microcanonical simulations the BHMC gives results for a very broad temperature range with computer efforts smaller than those required by five canonical Metropolis simulations.

1 Introduction

In order to calculate average values $<Q>_T$ of a certain quantity Q for a system in equilibrium at temperature T according to, for instance, the canonical ensemble

$$<Q>_T = \frac{\sum_E <Q>_E \, g(E) \exp\left(-E/k_B T\right)}{\sum_E g(E) \exp\left(-E/k_B T\right)}, \tag{1}$$

it is sufficient to know $g(E)$, that is, the number of states with energy E, and $<Q>_E$, which is the microcanonical average of Q at energy E. These data for $g(E)$ and $<Q>_E$, combined with the Boltzmann factor $\exp\left(-E/k_B T\right)$, give $<Q>_T$ at any desired temperature (in the following the Boltzmann constant $k_B = 1$ is taken).

2 The BHMC Method

The Broad Histogram method (BHMC), developed by de Oliveira et al. [3]-[6] calculates $g(E)$ and $<Q>_E$ directly. First, the method chooses a protocol of

Springer Proceedings in Physics, Volume 85
Computer Simulation Studies in Condensed-Matter Physics XII
Eds.: D.P. Landau, S.P. Lewis and H.-B. Schüttler

virtual random movements in the space of states of the system and calculates $N_{\text{up}}(X)$ ($N_{\text{dn}}(X)$) as the probability to increase (decrease) the energy of the configuration X by a fixed amount ΔE_{fix}. These movements are *virtual* in the sense that they are *never performed*; they are used to determine $g(E)$. The only condition imposed to the protocol is microreversibility (for discrete systems) or equal probabilities to perform a movement and to revert it (for continuous ones), that is,

$$P(X_{\text{old}} \to X_{\text{new}}) = P(X_{\text{new}} \to X_{\text{old}}) \,. \tag{2}$$

Due to this condition, it can be shown [6,7] that

$$g(E){<}N_{\text{up}}(E){>} = g(E+\Delta E_{\text{fix}}){<}N_{\text{dn}}(E+\Delta E_{\text{fix}}){>} \tag{3}$$

holds, where $<N_{\text{up}}(E)>$ ($<N_{\text{dn}}(E)>$) is the microcanonical average of $N_{\text{up}}(X)$ ($N_{\text{dn}}(X)$) at energy E. For continuous systems

$$N_{\text{up}}(X) \equiv f^X_{\Delta E}(\Delta E_{\text{fix}}) \,; \quad N_{\text{dn}}(X) \equiv f^X_{\Delta E}(-\Delta E_{\text{fix}}) \,, \tag{4}$$

where $f^X_{\Delta E}(\Delta E)$ denotes the probability of obtaining an energy change between ΔE and $\Delta E + d\Delta E$ starting from the configuration X, that is, the probability density function of ΔE starting from X.

Equation (3) can be rewritten as

$$\begin{aligned}\beta(E) \equiv \frac{d\ln g(E)}{dE} &\simeq \frac{\ln g(E+\Delta E_{\text{fix}}) - \ln g(E)}{\Delta E_{\text{fix}}} \\ &\simeq \frac{1}{\Delta E_{\text{fix}}} \ln \frac{<N_{\text{up}}(E)>}{<N_{\text{dn}}(E+\Delta E_{\text{fix}})>} .\end{aligned} \tag{5}$$

The function $\log g(E)$ is obtained either by adding differences or by integration of $\beta(E)$.

Next, a Monte Carlo sampling strategy to take the samples is chosen. Many different sampling strategies can be used in conjunction with the BHMC analysis technique. Because we are interested in the microcanonical averages $<N_{\text{up}}(E)>$, $<N_{\text{dn}}(E)>$ and $<Q(E)>$, the only requirement here is that all configurations with the same energy have to be sampled with the same probability. Therefore, the sampling dynamics must hold a detailed balance condition with probabilities depending on the energy only [8]

$$P(A)W(A \to B) = P(B)W(B \to A); \qquad P(A) = P(E(A)) \,, \tag{6}$$

where A, B are configurations. These movements are *real*, in the sense that they *are performed*. Several Monte Carlo procedures fulfill this condition, for instance, microcanonical [5], canonical and multicanonical simulations [9], and also a special kind of random walk described in [8]. All of them can be used in conjunction with the BHMC to estimate $g(E)$ exactly.

3 Example: The Classical Heisenberg Model

Let us choose the classical Heisenberg model with Hamiltonian

$$\mathcal{H} = -J \sum_{<ij>} \boldsymbol{\sigma}_i \cdot \boldsymbol{\sigma}_j \ , \tag{7}$$

where the summations $<ij>$ are taken over all pairs of nearest-neighbor sites and J is the maximal energy per bond ($J = 1$ is taken in the following).

First, let us choose the following protocol of virtual random movements: we take one spin at random and turn it to a new orientation homogeneously distributed on the 3D-sphere of radius one. Clearly, this protocol fulfills (2).

To find $f^X_{\Delta E}$ let us look at the energy of one spin $\boldsymbol{\sigma}_i$, and define

$$\varepsilon_i \equiv -\sum_j \boldsymbol{\sigma}_i \cdot \boldsymbol{\sigma}_j = -\boldsymbol{\sigma}_i \cdot \boldsymbol{A}_i \quad ; \quad \boldsymbol{A}_i \equiv \sum_j \boldsymbol{\sigma}_j \ , \tag{8}$$

where the summations j are taken over the nearest-neighbors of site i. Let (θ, φ) be the coordinates of the new orientation of $\boldsymbol{\sigma}_i$ in a spherical coordinate system with the z axis parallel to $\boldsymbol{A}_i$. An homogeneous distribution on the sphere is described by

$$f_{\theta,\varphi}(\theta, \varphi) = \frac{1}{4\pi} \sin\theta; \qquad \theta \in (0, \pi) \ , \ \varphi \in (0, 2\pi) \ . \tag{9}$$

It is clear that $\varepsilon_i = -A_i \cos\theta$. Let us take φ as an auxiliary variable and calculate $f_{\varepsilon_i,\varphi}$. The rules for functions of random variables [10] give

$$f_{\varepsilon_i,\varphi}(\varepsilon_i, \varphi) = \frac{f_{\theta,\varphi}(\theta_1, \varphi_1)}{|J(\theta_1, \varphi_1)|} \quad ; \quad \begin{array}{l} \theta_1 = \cos^{-1}(-\varepsilon_i/A_i) \\ \varphi_1 = \varphi \end{array} \ . \tag{10}$$

$J(\theta, \varphi) \equiv \partial(\varepsilon_i, \varphi)/\partial(\theta, \varphi) = A_i \sin(\theta)$ is the Jacobian of this transformation. It follows that

$$f_{\varepsilon_i,\varphi}(\varepsilon, \varphi) = \frac{1}{4\pi A_i} \quad ; \quad f_{\varepsilon_i}(\varepsilon_i) = \int_{\varphi=0}^{2\pi} f_{\varepsilon_i,\varphi}(\varepsilon, \varphi) d\varphi = \frac{1}{2A_i} \ . \tag{11}$$

Because $\varepsilon_i = \Delta E + \varepsilon_{i_{\text{old}}}$ holds, we can write

$$f^i_{\Delta E}(\Delta E) = \left\{ \begin{array}{ll} \frac{1}{2A_i} & ; \ \text{if } |\Delta E + \varepsilon_{i\text{old}}| < A_i \\ 0 & ; \ \text{otherwise} \end{array} \right\} \ . \tag{12}$$

Finally, $f^X_{\Delta E} = \frac{1}{N} \sum_i f^i_{\Delta E}$, with N the number of sites, and $N_{\text{up}}(X)$ ($N_{\text{dn}}(X)$) equals the average of $f^i(\Delta E_{\text{fix}})$ ($f^i(-\Delta E_{\text{fix}})$) on all sites.

Next, let us choose the samples by using some Monte Carlo strategy. We took a set of energy values and performed a microcanonical simulation at each one as follows. We took an energy window of width E_b centered at the energy of interest E_o, pick a spin at random and choose a new orientation for the spin according to

$$f_{\theta,\varphi}(\theta, \varphi) \propto \sin\theta; \qquad \theta \in (\theta_{\min}, \theta_{\max}) \ , \ \varphi \in (0, 2\pi) \ , \tag{13}$$

where $\theta_{\min}$ and $\theta_{\max}$ are defined so that the new energy falls always inside the window. Furthermore, (13) ensures detailed balance condition. To estimate $\beta(E)$ we used the approximation $<N_{\mathrm{dn}}(E_o + \Delta E_{\mathrm{fix}})> \simeq <N_{\mathrm{dn}}(E_o)>$ [11], that is, by both microcanonical averages in (5) were taken from the same microcanonical simulation.

4 Results

Figure 1 (**a**) shows $\beta(E)$ obtained from the BHMC method for the SC $10 \times 10 \times 10$ classical Heisenberg model with periodic boundary conditions, by performing microcanonical simulations at 120 equally spaced energy values on the range of energies per bond between -0.9 and 0.0. We used $\Delta E_{\mathrm{fix}} = 2.5$ and $E_b = 3.0$. The error bars (not shown) are of the order of 0.1%. Figure 1 (**b**) compares the specific heat obtained from this data with that obtained from Metropolis simulations. The agreement between the two methods is excellent. Furthermore, the BHMC method took 282 s per run on an Alpha workstation and each Metropolis simulation took 62 s on the same machine. It means, that the BHMC calculates the whole temperature range $0.5 < T < 3.0$ by using less CPU time than five temperature points using Metropolis. Further details will be presented elsewhere.

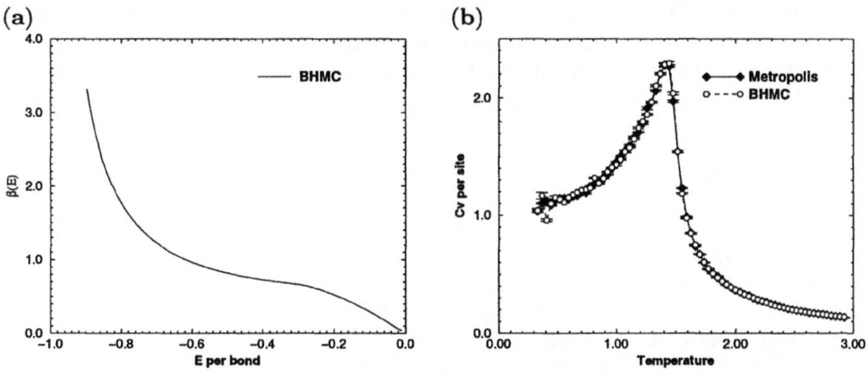

Fig. 1. $\beta(E)$ (**a**) and heat capacity (**b**) for the SC $10 \times 10 \times 10$ classical Heisenberg model obtained from the BHMC method (*circles*) and canonical Metropolis simulations (*filled diamonds*)

5 How to Include an External Field

Let us consider a system with a Hamiltonian

$$\mathcal{H} = -J \sum_{<ij>} \boldsymbol{\sigma}_i \cdot \boldsymbol{\sigma}_j - \boldsymbol{H} \cdot \boldsymbol{\sigma}_i \, , \tag{14}$$

where $\boldsymbol{H}$ is the external field. The spins can rotate either in a plane (XY-model) or in the 3D-space (Heisenberg model). Let us look at the energy of one spin, defined by

$$\varepsilon_i \equiv -\sum_j \boldsymbol{\sigma}_i \cdot \boldsymbol{\sigma}_j - \boldsymbol{H} \cdot \boldsymbol{\sigma}_i = -\boldsymbol{\sigma}_i \cdot \boldsymbol{A}_i^* \;\; ; \; \boldsymbol{A}_i^* \equiv \boldsymbol{A}_i + \boldsymbol{H} \,. \tag{15}$$

Therefore, all we need to add an external field is to replace $\boldsymbol{A}_i$ by $\boldsymbol{A}_i^*$ and use the same procedure to find $f_{\Delta E}^X$ as before. The same strategy can be used to include long-range interactions. Results for the XY-model with external field will be presented elsewhere.

6 Conclusions

In conclusion, the BHMC method gives excellent results for another continuous system, namely, the classical Heisenberg model. In addition, it is easy to extend the method for systems with an external field. These results support the generality of the method.

7 Acknowledgments

We thank P.M.C. de Oliveira, D. Stauffer, B. Berg and W. Janke for helpful discussions. We are also grateful to H. Gould and F. Kun for valuable suggestions on earlier version of the manuscript. The work of one of the authors (Muñoz) is supported by the Deutscher Akademischer Austauschdienst (DAAD) through scholarship A/96/0390.

References

1. J.D. Muñoz and H.J. Herrmann, *Int. J. Mod. Phys. C* (in print), also in cond-mat/9810024.
2. J.D. Muñoz and H.J. Herrmann, cond-mat/9810292.
3. P.M.C. de Oliveira, T.J.P. Penna and H.J. Herrmann, *Braz. J. of Phys.* **26**, 677-683 (1996); also in cond-mat/9610041.
4. P.M.C. de Oliveira, T.J.P. Penna and H.J. Herrmann, *Eur. Phys. J. B* **1**, 205-208 (1998); also in cond-mat/9709064.
5. P.M.C. de Oliveira, *Int. J. Mod. Phys. C* **9**, 497-503 (1998).
6. P.M.C. de Oliveira, *Eur. Phys. J. B* **6**, 111-115 (1998).
7. B.A. Berg and U.H.E. Hansmann, *Eur. Phys. J. B* **6**, 395-398 (1998).
8. J.-S. Wang, cond-mat/9810017, to appear in *Eur. Phys. J. B.*
9. B.A. Berg, *Int. J. Mod. Phys. C* **3**, 1083-1098 (1992).
10. A. Papoulis, *Probability, Random Variables, and Stochastic Processes* (McGraw-Hill, New York, 1965) Chaps. 5-7.
11. The function $\ln(<N_{dn}(E)>)$ changes slowly on the whole energy axis except at very low energies, which do not affect the temperatures we are interested in. Furthermore, we are taking E_b = 0.1% of the total energy. This justifies the approximation.

Introduction to the Propp-Wilson Method of Exact Sampling for the Ising Model

M. A. Novotny

Supercomputer Computations Research Institute, Florida State University, Tallahassee, FL 32306-4130, USA

Abstract. An introduction to the Propp-Wilson method of coupling-from-the-past for the Ising model is presented. It enables one to obtain exact samples from the equilibrium spin distribution for ferromagnetic interactions. Both uniform and random quenched magnetic fields are included. The time to couple from the past and its standard deviation are shown as functions of system size, temperature, and the field strength. The time should be an estimate of the ergodic time for the Ising system.

The standard importance-sampling Monte Carlo algorithm [1] for statics has a number of difficulties. One is its inability to ensure that the generated spin configurations are uncorrelated and are representative of the equilibrium distribution. The standard way to overcome this difficulty is to throw away a large number of Monte Carlo steps between analyzed configurations. Of course this decreases the number of configurations obtained in a given amount of computer time. In addition, there can remain the question of whether the generated spin configurations are indeed drawn from the underlying equilibrium distribution. For example, consider a spin-glass model at low temperatures, where ergodicity has been broken for reasonable simulation times and the system might have many ground states and many more metastable states. Then it is extremely difficult to decide whether the standard Monte Carlo method obtains spin configurations drawn from the underlying equilibrium spin distribution.

In this short paper the square-lattice Ising model with nearest-neighbor ferromagnetic ($J>0$) interactions and periodic boundary conditions is considered. The Hamiltonian is $\mathcal{H} = -J\sum_{\langle ij \rangle} \sigma_i\sigma_j - \sum_i H_i\sigma_i$, with $\sigma_i=\pm 1$, the first sum over nearest-neighbor spins, and H_i is the local field. Two cases are considered, uniform field $H_i=H$, and quenched random field with H_i chosen to be $\pm H$ with equal probability. We take $J=1$, which makes the exact critical temperature $T_c \approx 2.26\cdots$.

One Monte Carlo step consists of two procedures. First randomly choose a spin i from the $L\times L$ lattice. The probability of σ_i being set to $+1$ in the next time step is

$$\mathrm{Prob}(\sigma_i = 1) = \left\{ 1 + \exp\left[\left(-2J \sum_{j=\mathrm{nn}} \sigma_j - 2H_i \right) \Big/ T \right] \right\}^{-1} , \tag{1}$$

Springer Proceedings in Physics, Volume 85
Computer Simulation Studies in Condensed-Matter Physics XII
Eds.: D.P. Landau, S.P. Lewis and H.-B. Schüttler

where the first sum is over the nearest-neighbors of spin i and T is the temperature. The second step is to use a uniformly distributed random number r, and set σ_i=+1 in the next time step if $r \leq \text{Prob}(\sigma_i = 1)$; otherwise set σ_i=−1. This is the single-spin flip heat-bath dynamic, which for the Ising model is the Glauber dynamic. I use the Ziff four-tap random number generator [2]. It is important to note that we do *not* consider the probability that a spin flips, but rather the probability of the spin being up, *i.e.* +1.

Recently researchers in the statistics community have devised methods that can guarantee that generated spin configurations from a Monte Carlo simulation are drawn from the underlying equilibrium distribution. There the study goes under the name of Markov random fields [3], which should *not* be confused with the random fields in the Hamiltonian. The method used here is the 1996 method of Propp and Wilson [4]. It is referred to as 'coupling-from-the-past' or the 'Propp-Wilson method'. Here we bring the single-spin-flip version of the Propp-Wilson method to the attention of practitioners in the statistical mechanical community. We obtain a time for coupling from the past, which we call the Propp-Wilson time, t_{PW}. We will show how t_{PW} changes with T, L, H, and the randomness in H. The time t_{PW} can be viewed as an upper bound to the ergodic time τ_{e} of the simulation [1]. In the spirit of a workshop, this study was only initiated about one month prior to the workshop. Consequently, it should be viewed as an exploratory and preliminary work.

The basic idea of the Propp-Wilson method is to not run the Markov chain forward in time (which all normal importance-sampling Monte Carlo procedures do), but rather to run the Markov chain from a time $-t$ in the past up to t=0. Typically it is not known how to go from a spin configuration at one time to one at an earlier time. However, given a spin configuration at time $-t$, it is easy to propagate it to time t=0 — just use the algorithm for forward propagation. What is the difference between going from $-t$ to 0 or going from 0 to t? It lies only in our knowledge of the sequence of random numbers used. At t=0 one has no idea which spin will be picked next or which random number the spin-up probability will be compared with. However, at all times in the past, both the spin picked and the random number to compare with are known.

Now consider starting two spin configurations at time $-t$, and propagating forward in time toward t=0. The *same* random numbers are used for each spin configuration, since the same history of random numbers applies for times $t \leq 0$. At some time $-t_{\text{c}}$ these two spin configurations might coalesce, i.e. they both become the same spin configuration at time $-t_{\text{c}}$. From this time onward to time t=0 they will remain coalesced, since the same random numbers propagate them forward to t=0. This coalescence gives the Propp-Wilson method the 'coupling from the past' name.

Here is the Propp-Wilson prescription for simulation of N Ising spins. First obtain a *very long* history of randomly picked spins and random numbers used to compare against $\mathrm{Prob}(\sigma_i = +1)$. Start with all 2^N possible spin configurations at time $t{=}-1$. Propagate all configurations forward in time to $t{=}0$. Remember, this is done using the *same* set of random numbers for each configuration. If *all* 2^N configurations have coalesced to the same spin configuration at $t{=}0$, this is the spin configuration you want. If not, start all 2^N spin configurations at $t{=}-2$, propagate forward with the same random number history, and see if they have coalesced at $t{=}0$. This is continued until at some time $-t_{\mathrm{PW}}$ the coalescence of all 2^N spin configurations has occurred. The spin configuration obtained in this fashion is *guaranteed* to be an exact sampling of the equilibrium spin distribution. An exact sample means that the spin configuration is drawn according to the probability associated with the equilibrium distribution. Note that care must be taken to ensure that the long random-number history is never exceeded before the coalescence occurs.

Why does this give an exact sampling? Let us imagine that we have run an infinite ensemble of spin configurations from $t{=}-\infty$ to $t{=}-t_{\mathrm{PW}}$ using different random numbers for each. Since we have run our Markov chain infinitely long, we are guaranteed that the spin configurations at $t{=}-t_{\mathrm{PW}}$ are distributed according to the equilibrium spin distribution. Each of the starting spin configurations will have evolved into one of the 2^N spin configurations at $t{=}-t_{\mathrm{PW}}$. No matter which spin configuration the system is in at time $-t_{\mathrm{PW}}$, we know that at time $t{=}0$ all 2^N configurations have coalesced to a single spin configuration since we have used our history of random numbers which is the same for each configuration to run from $t{=}-t_{\mathrm{PW}}$ to $t{=}0$. Thus the sample generated is drawn from the equilibrium spin distribution.

In principle only a coupling-from-the-past method, like the Propp-Wilson method, is required to obtain a perfectly random sample. In practice another relationship is required since the number of states grows very quickly with lattice size. If all 2^N states could be handled, the partition function could be calculated exactly. For special cases it is possible to introduce a **partial ordering** on the states. For the Ising model, we say that two spin configurations $\boldsymbol{x}$ and $\boldsymbol{y}$ have a partial ordering, $\boldsymbol{x} \preceq \boldsymbol{y}$, if for each lattice site the spin in $\boldsymbol{x}$ is down whenever the spin in $\boldsymbol{y}$ is down. The transition probability of (1) preserves the partial ordering for ferromagnetic interactions. Consequently, rather than insist that all 2^N configurations have coalesced, only the two extreme partial ordering states, all spins up and all spins down, need to be considered. If these two states have coalesced, then all other states have coalesced. It is this desired property of partial ordering that restricts one to ferromagnetic interactions.

The Propp-Wilson method was run to obtain t_{PW} as a function of L, T, and H. For each value of parameters 1000 exact samples were obtained. Note that if coalescence occurs at time $-t_{\mathrm{PW}}$, it also occurs for any earlier time. This allows a bisection method to be used to find t_{PW}. Figure 1(a) shows the average value of the Propp-Wilson time, $[t_{\mathrm{PW}}]$, in Monte Carlo Steps per Spin (MCSS), at $H{=}0$. Since the phase space grows with system size, $[t_{\mathrm{PW}}]$

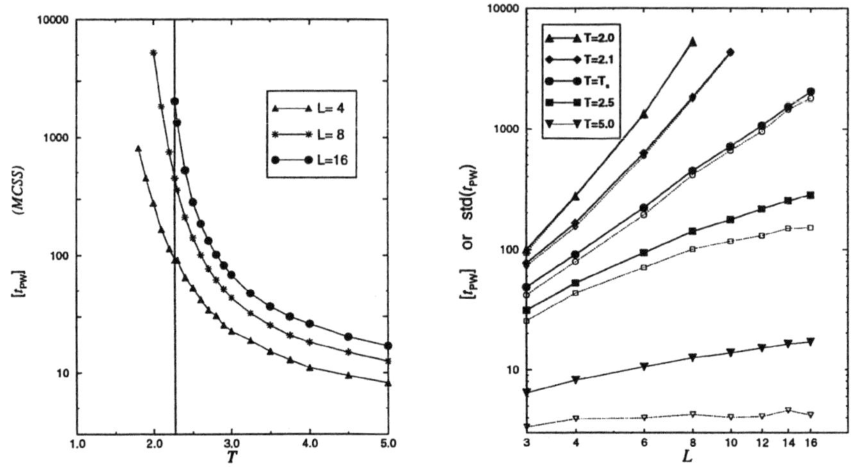

Fig. 1. (a) The average Propp-Wilson time, $[t_{\rm PW}]$, is shown as a function of temperature for H=0 and L=4, 8, and 16. The vertical line is the exact value of the critical temperature, $T_{\rm c}$. The lines are guides for the eye. (b) The L dependence on a log-log plot is shown for $[t_{\rm PW}]$ (filled symbols joined by solid lines) and the standard deviation of $t_{\rm PW}$ (corresponding open symbols joined by dashed lines). Both of these grow faster than power laws below $T_{\rm c}$, slower than power laws above $T_{\rm c}$, and are expected to grow as a power law at $T_{\rm c}$

increases with L. Furthermore, since it takes longer to sample all of phase space at lower temperatures, $[t_{\rm PW}]$ increases as T is lowered.

For a d dimensional system with correlation length ξ, below but near $T_{\rm c}$ the ergodic time is expected to scale as $\tau_{\rm e} \propto L^z \exp\left[{\rm const}(L/\xi)^{d-1}\right]$ [1]. Here z is the exponent associated with critical slowing down. For $T \geq T_{\rm c}$ the 'intrinsic' order parameter relaxation time τ is the longest relaxation time in the system [1], so above $T_{\rm c}$ one expects $\tau \propto \xi^z \propto |1 - T/T_{\rm c}|^{-\nu z}$ where ν is the correlation length exponent. Right at $T_{\rm c}$ this yields $\tau \propto L^z$. Figure 1(b) shows the L dependence of $[t_{\rm PW}]$ for H=0 and various temperatures. The graph illustrates that this time grows slower (faster) than a power law above (below) $T_{\rm c}$. At $T_{\rm c}$ this time seems to scale as a power law. However, care must be used in associating $[t_{\rm PW}]$ with either τ or $\tau_{\rm e}$. One might expect that $[t_{\rm PW}]$ yields an upper bound on the larger of $\tau_{\rm e}$ or τ. As seen in Fig. 1 even at high temperatures $[t_{\rm PW}]$ increases with L. These additional size dependent factors in $[t_{\rm PW}]$ might also be expected to be present at lower temperatures. In a recent thesis [5] $[t_{\rm PW}] \propto L^{\Omega}$ was postulated, and the value $\Omega \approx 2.44$ was found with system sizes L up to 310. (This assumes that the presented values are at $T_{\rm c}$, which was not stated explicitly.) This is somewhat high compared to the accepted value of $z \approx 2.21$ [6]. It is also possible to calculate higher moments of $t_{\rm PW}$. Figure 1(b) shows the standard deviation per spin for $t_{\rm PW}$

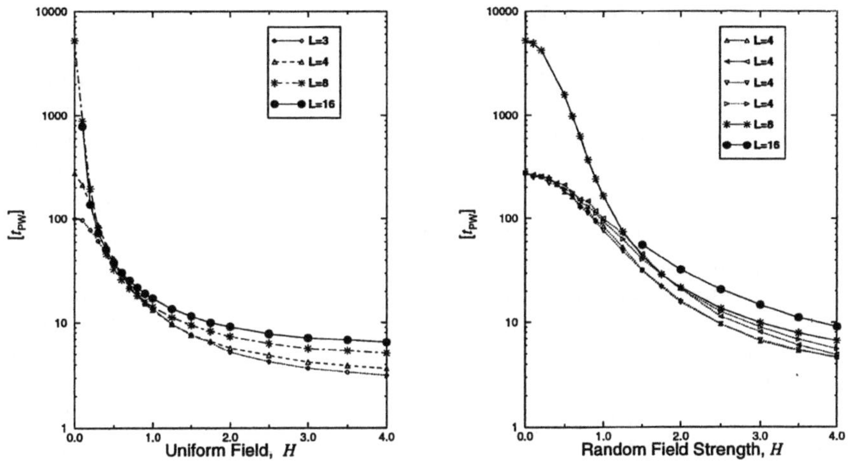

Fig. 2. The average Propp-Wilson time, $[t_{\mathrm{PW}}]$, in MCSS is shown at T=2.0 as a function of H. The lines are guides for the eye. Note that the points at H=0 are the same for both plots. (a) Is for a uniform magnetic field. For L=16, H=0.1 is the smallest value shown. (b) Is for quenched randomly distributed fields equal to $-H$ or $+H$. Four different distributions of fields are shown for L=4, and one distribution for L=8 and L=16. For L=16, H=1.5 is the smallest value shown

as a function of L. The behavior seems to be close to those of $[t_{\mathrm{PW}}]$. Note that the standard deviation becomes very large, which necessitates using a large number of samples to obtain a reasonable value for $[t_{\mathrm{PW}}]$.

Figure 2(a) shows $[t_{\mathrm{PW}}]$ at T=2.0 for different values of L and uniform applied field, H. At strong fields the larger system sizes have larger values of $[t_{\mathrm{PW}}]$, for intermediate fields all system sizes have similar values of $[t_{\mathrm{PW}}]$, and for weak fields the smaller system sizes break away from this curve and go to their maximum at H=0. This behavior is reminiscent of the cross-overs encountered in the study of metastability, where for $T<T_{\mathrm{c}}$ as the field is decreased different regimes are encountered in a finite system [7].

To illustrate the application of the Propp-Wilson method to a random-field Ising model, Fig. 2(b) shows data for $[t_{\mathrm{PW}}]$ as a function of the strength of the random field for various quenched field configurations for L=4, and one quenched field configuration for L=8 and 16. The trends seem to be similar to those of Fig. 2(a). However the values for $[t_{\mathrm{PW}}]$ seem to increase faster as H decreases than in the uniform field case. This is in agreement with the physical idea that a system with random quenched disorder has a longer ergodicity time than one without disorder. For more discussion of random fields see references [8–10].

In summary, it is possible to utilize coupling-from-the-past ideas to obtain perfect samples of Ising spin configurations from the equilibrium spin distribution. To use the ideas in practice, however, another relationship is essential, that of a partial ordering preserved by the transition probabilities. This restriction allows one to study random fields, but no partial ordering is known for systems with antiferromagnetic bonds. The value for the average and standard deviation for t_{PW} seem to have reasonable dependences on system size, field strength, and temperature if t_{PW} is viewed as an upper bound for the ergodic time of the system.

It is possible to use cluster dynamics instead of single-spin-flip dynamics. Since this decreases the ergodic time near T_c, this allows one to obtain perfect samples for large systems — so far the largest single example I have seen had L=2100 in a preprint by Propp and Wilson. It has also been shown that the expected running time for the Propp-Wilson method should be within a constant multiple of the maximum expected time for two states to coalesce in the Markov chain, and that this running time is near optimal for exact sampling. Thus the ability to obtain results for large systems at temperatures above and near T_c for models that have a partial ordering seems realistic. Unfortunately, these partial orderings are extremely difficult to find in more general models. Also, the long expected run times seem to make the application of these types of algorithms rather limited in studies of low-temperature behavior for systems with quenched randomness and high barriers.

This work is supported in part by the NSF through grant number 9871455, and through the Supercomputer Computations Research Institute which is funded by the U.S. DoE and the State of Florida.

References

1. Binder K. and Heermann D. W. (1997) *Monte Carlo Simulation in Statistical Physics*, third edition, Springer, Berlin
2. Ziff R.M. (1998) Computers in Physics **12**, 385
3. Kindermann R. and Snell J. L. (1980) *Markov Random Fields and their Applications*, American Mathematical Society, Providence, R.I.
4. Propp J. G. and Wilson D. B. (1996) Random Structures and Algorithms **9**, 223
5. Fismen M. (1998) *Exact Simulation using Markov Chains*, diploma thesis, Dept. of Mathematics, The Norwegian University of Science and Technology, Trondheim, Norway
6. Wang J. S. (1993) Phys. Rev. B **47**, 869
7. Rikvold P.A., Novotny M.A., Kolesik M., and Richards H.L. (1998) In *Dynamical Properties of Unconventional Magnetic Systems*, editors Skjeltorp A.T. and Sherrington D., Kluwer, Dordrecht, 307
8. Birgeneau R. J. et al (1995) Phys. Rev. Lett. **75**, 1198
9. Alava M. and Rieger H. (1998) Phys. Rev. E **58**, 4284; Rasmussen E. B., Novotny M. A., and Landau D. P. (1982) J. Appl. Phys. **53**, 1925
10. Jolicœur T. and Le Guillou J.-C. (1997) Phys. Rev. B **56**, 10766; Novotny M. A. (1992) Phys. Rev. B **46**, 2939

Magnetic Properties of Finite Systems: Microcanonical Finite-Size Scaling

M. Promberger, M. Kastner, and A. Hüller

Institut für Theoretische Physik, Staudtstraße 7B3, D–91058 Erlangen, Germany

Abstract. In the microcanonical ensemble, suitably defined observables show non-analyticities and power law behaviour even for finite systems. For these observables, a microcanonical finite-size scaling theory is established which facilitates an approach to the critical exponents of the infinite system.

1 Introduction: Microcanonical Description of Finite Systems

In a microcanonical description of finite systems, the magnetic equation of state and the zero-field isothermal magnetic susceptibility are defined in terms of derivatives of the microcanonical entropy $s(\varepsilon,m,L^{-1})$ (cf. [2],[3]):

$$m_{h=0}(\varepsilon,L^{-1}) \quad \Longleftarrow \quad \max_{m\geq 0}\{s(\varepsilon,m,L^{-1})\} \;=\; s\left(\varepsilon,m_{h=0}(\varepsilon,L^{-1}),L^{-1}\right) \tag{1}$$

$$\chi_{T;h=0}(\varepsilon,L^{-1}) \;:=\; \left.\frac{\partial m}{\partial h}\right|_{T;h=0}(\varepsilon,L^{-1}) \;= \tag{2}$$

$$= \left\{\left[\frac{\partial s}{\partial \varepsilon}\left[\left(\frac{\partial^2 s}{\partial\varepsilon\partial m}\right)^2 \Big/ \frac{\partial^2 s}{\partial\varepsilon^2} - \frac{\partial^2 s}{\partial m^2}\right]^{-1}\right](\varepsilon,m,L^{-1})\right\}_{m=m_{h=0}(\varepsilon,L^{-1})}$$

Here, ε denotes the specific enthalpy (cf. [1]), h the magnetic field, m the specific magnetization, L the linear system size and d the spatial dimension. For notational convenience, the zero field limit is denoted by $h=0$. By definition, the microcanonical entropy is the logarithm of the microcanonical partition function (density of states) Ω:

$$s(\varepsilon,m,L^{-1}) \;=\; L^{-d}\ln\Omega(\varepsilon,m,L^{-1}) \quad . \tag{3}$$

In Fig. 1, the zero-field magnetization (Fig. 1a) and the zero-field magnetic susceptibility (Fig. 1b) are plotted for the example of the $3d$-Ising system for various system sizes. Both observables show a behaviour reminiscent of the critical behaviour of the infinite system and therefore suggest the introduction of finite system exponents $\beta_{\varepsilon,L}$ and $\gamma_{\varepsilon,L}$ (describing the power law behaviour with respect to the base $|\tilde{\varepsilon}|$ defined below; cf. Ref. 13 of [2]).

Springer Proceedings in Physics, Volume 85
Computer Simulation Studies in Condensed-Matter Physics XII
Eds.: D.P. Landau, S.P. Lewis and H.-B. Schüttler

The transition energies $\varepsilon_T(L)$, at which the non-analyticities occur, are obviously dependent on system size.
In Figs. 1c/d, magnetizations and susceptibilities are plotted as functions of the reduced enthalpy $\tilde{\varepsilon}(L) := (\varepsilon - \varepsilon_T(L))/|\varepsilon_T(L)|$.

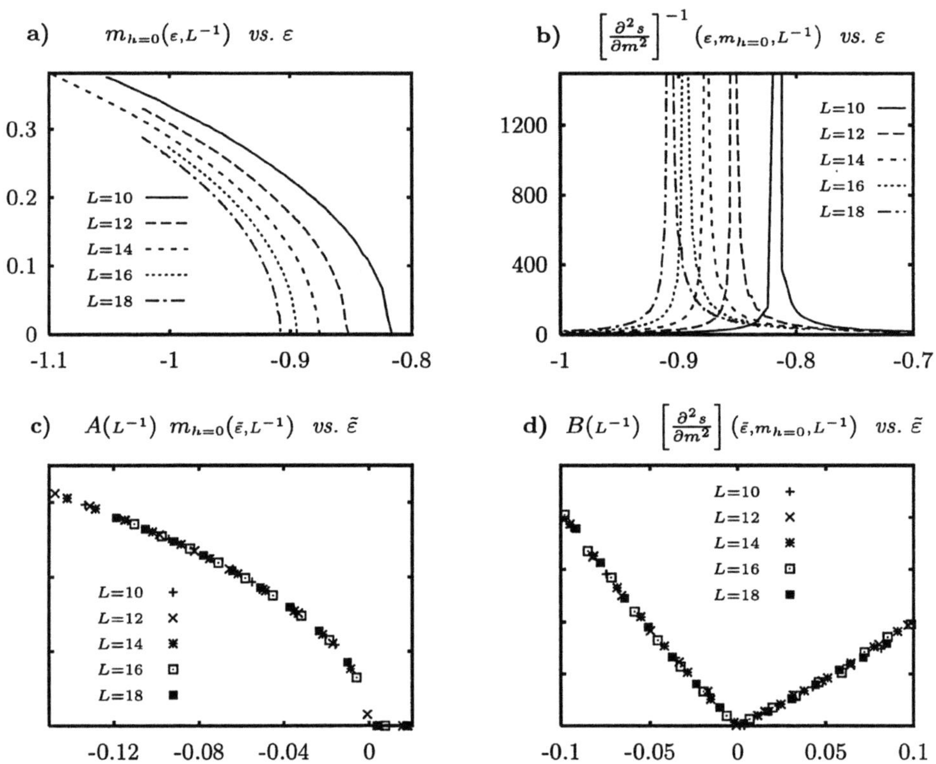

Fig. 1. Microcanonical magnetizations and "susceptibilities" of finite $3d$-Ising systems (Figs. a/b). For numerical convenience, $\left[(\partial^2 s)/(\partial m^2)\right]^{-1}$ is shown instead of $\chi_{T;h=0}$. Note that both quantities show the same microcanonical finite-size scaling behaviour and – in the thermodynamic limit – the same critical behaviour. Multiplication of $m_{h=0}$ and $(\partial^2 s)/(\partial m^2)$ by suitable scaling factors $A(L^{-1})$ and $B(L^{-1})$ and plotting the thus obtained results against $\tilde{\varepsilon}$ yields data collapse (Figs. c/d). Whereas Figs. a/b suggest the introduction of finite system exponents, Figs. c/d give rise to the assumption that the thus defined exponents show no system size dependence

Data collapse is achieved by suitable scaling of the amplitudes of $m_{h=0}$ and $\chi_{T;h=0}$. Therefore, it seems to be evident that the finite system critical exponents are independent of the system size (i.e. take on the same value $\forall L^{-1} \neq 0$), but differ from the values expected in the thermodynamic limit (**TDL**), $\beta_{\varepsilon,\infty} \approx 0.37$, $\gamma_{\varepsilon,\infty} \approx 1.38$, whereas from numerical data we obtain $\beta_{\varepsilon,L} \approx 0.5$, $\gamma_{\varepsilon,L} \approx 1$ for all L considered. In this paper, a microcanonical finite size scaling (**MFSS**) theory is developed, taking into account the following constraints (partly justified above):

(i) consistence with the canonical finite size scaling (**CFSS**) theory

(ii) power law behaviour of the finite system magnetization and susceptibility

(iii) finite system critical exponents which do not depend on the system size and which are not necessarily identical to those of the infinite system

2 Microcanonical Finite-Size Scaling (MFSS)

In the vicinity of a critical point $(t, h, \varepsilon^*, m \approx 0)$, the thermodynamic potentials can be divided into a singular part (superscript *) which describes the non-analytic behaviour, and a regular part (subscript $_{reg}$). For the free energy density and the specific entropy this reads:

$$g(t,h) = g^*(t,h) + g_{reg}(t,h) \quad , \quad t{:=}\tfrac{T-T_c}{T_c} \quad , \quad \begin{matrix} T_c \text{ critical temperature} \\ h \text{ magnetic field} \end{matrix} \tag{4}$$

$$s(\varepsilon^*,m) = s^*(\varepsilon^*,m) + s_{reg}(\varepsilon^*,m) \quad , \quad \varepsilon^*{:=}\tfrac{\varepsilon-\varepsilon_c}{|\varepsilon_c|} \quad , \quad \varepsilon_c \text{ critical enthalpy} \tag{5}$$

It can be shown (see e.g. [4] and references therein) that in the TDL the singular parts of the various potentials are homogeneous functions (cf. Eqs. (6),(7) in Fig. 2) and all critical exponents can be expressed in terms of the degrees of homogeneity ($a_\varepsilon = (1-\alpha)/(2-\alpha)$, $a_m = 1/(\delta+1)$, $a_t = 1 - a_\varepsilon$, $a_h = 1 - a_m$). The equivalence of ensembles is valid only in the TDL. Hence, the thermodynamic potentials of finite systems have to be classified as canonical or microcanonical quantities (cf. [2],[6]). Starting point for the CFSS is the so-called "CFSS assumption" [7]: For large but finite systems, $g_c^*(t,h,L^{-1})$ is a homogeneous function in accordance with Eq. (8).
We have shown elsewhere [6] that the MFSS assumption (9) entails (8), whereas proof of the reverse is more difficult. Note that Eq. (9) even accounts for the possibility of s_m^* showing no system size dependence at all; nevertheless, this case results in an L-dependence of g_c^* and in CFSS. In the rest of this paper, we will discuss the consequences of the MFSS assumption which states that:

MFSS assumption : $s_m^*(\varepsilon^*,m,L^{-1})$ is a homogeneous function of its arguments (cf. Fig. 2, Eq. (9))

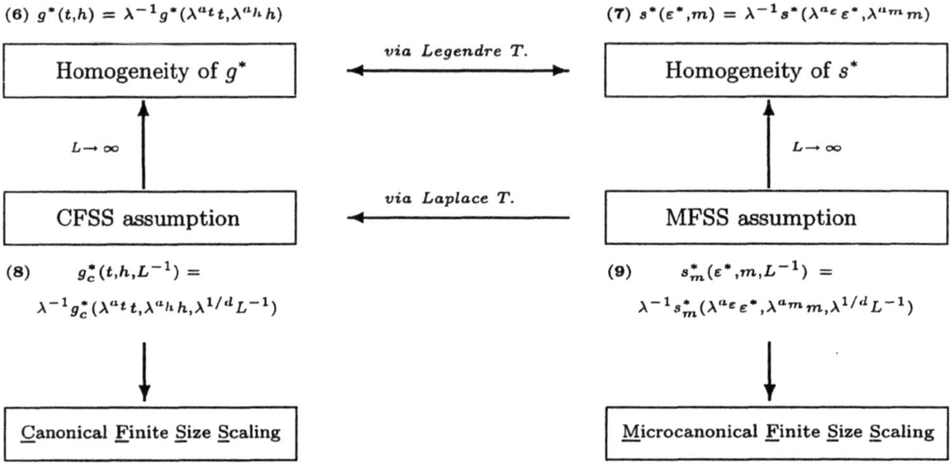

Fig. 2. Homogeneity relations (valid $\forall\lambda > 0$) for the singular parts of the free energy and the entropy for finite (Eqs. (8),(9)) as well as infinite systems (Eqs. (6),(7)). In the TDL, a Legendre transform connects the homogeneity of g^* to the homogeneity of s^* (see Ref. [5]), whereas for finite systems it can be shown that the CFSS assumption is a consequence of the MFSS assumption, i.e., MFSS is consistent with CFSS [6]

From the MFSS assumption (9) and Eqs. (1) and (2), the *MFSS relations of the magnetization and the susceptibility* are derived easily [3]:

$$m^*_{h=0}(\varepsilon^*,L^{-1}) = \lambda^{-a_m}m^*_{h=0}(\lambda^{a_\varepsilon}\varepsilon^*,\lambda^{1/d}L^{-1}) \overset{\lambda=L^d}{=} L^{-da_m}\Phi_{m^*}(L^{da_\varepsilon}\varepsilon^*) \quad (10)$$

$$\chi^*_{T;h=0}(\varepsilon^*,L^{-1}) = \lambda^{1-2a_m}\chi^*_{T;h=0}(\lambda^{a_\varepsilon}\varepsilon^*,\lambda^{1/d}L^{-1}) \overset{\lambda=L^d}{=} L^{d(1-2a_m)}\Phi_{\chi^*}(L^{da_\varepsilon}\varepsilon^*) \quad (11)$$

where the Φ_i are so-called MFSS functions which describe the behaviour of the magnetization and the susceptibility of finite systems in the vicinity of the critical point ε_c of the infinite system.

MFSS of the transition point: From Figs. 1a/c, it can be deduced that the MFSS function of the magnetization $\Phi_m(x)$ is zero for $L^{da_\varepsilon}\varepsilon^* =: x \geq x_T$, where x_T determines the magnetic transition point $\varepsilon_T(L)$ for finite systems:

$$x_T = L^{da_\varepsilon}\frac{\varepsilon_T(L) - \varepsilon_c}{|\varepsilon_c|} \iff \varepsilon_T(L) = \varepsilon_c(1+\tilde{x}_T L^{-da_\varepsilon}),\ \tilde{x}_T := \frac{\varepsilon_c}{|\varepsilon_c|}x_T \quad (12)$$

Remarks: a) An additional quantity $\tilde{s}^*_m$ can be introduced, which is defined to be the singular part of the entropy, written in terms of the reduced enthalpy $\tilde{\varepsilon}(L) := (\varepsilon - \varepsilon_T(L))/|\varepsilon_T(L)|$.

Using Eq. (12), the thus defined entropy $\tilde{s}^*_m$ can be shown to possess the same degrees of homogeneity as s^*_m:

$$\tilde{s}^*_m(\tilde{\varepsilon},m,L^{-1}) = \lambda^{-1}\tilde{s}^*_m(\lambda^{a_\varepsilon}\tilde{\varepsilon},\lambda^{a_m}m,\lambda^{1/d}L^{-1}) \ , \ \text{where} \tag{13}$$

$$\tilde{s}^*_m(\tilde{\varepsilon},m,L^{-1}) := s^*_m(\varepsilon^*{=}\tilde{\varepsilon}{+}x_T L^{-da_\varepsilon},m,L^{-1}) \ .$$

Starting from expression (13), MFSS relations can be derived for the magnetization and the susceptibility. They illustrate the system size dependence of the behaviour of $m_{h=0}$ and $\chi_{T;h=0}$ in the vicinity of the transition point $\varepsilon_T(L)$ of the finite system:

$$\tilde{m}_{h=0}(\tilde{\varepsilon},L^{-1}) = \lambda^{-a_m}\tilde{m}_{h=0}(\lambda^{a_\varepsilon}\tilde{\varepsilon},\lambda^{1/d}L^{-1}) \overset{\lambda=L^d}{=} L^{-da_m}\Phi_{\tilde{m}}(L^{da_\varepsilon}\tilde{\varepsilon}) \tag{14}$$

$$\tilde{\chi}_{T;h=0}(\tilde{\varepsilon},L^{-1}) = \lambda^{1-2a_m}\tilde{\chi}_{T;h=0}(\lambda^{a_\varepsilon}\tilde{\varepsilon},\lambda^{1/d}L^{-1}) \overset{\lambda=L^d}{=} L^{d(1-2a_m)}\Phi_{\tilde{\chi}}(L^{da_\varepsilon}\tilde{\varepsilon}) \tag{15}$$

b) Numerical data suggest power law behaviour for both the magnetization and the susceptibility of finite systems. Therefore, the MFSS functions $\Phi_{\tilde{m}}$ and $\Phi_{\tilde{\chi}}$ have to be power laws, governed by the respective finite system exponents which are <u>not</u> determined by the degrees of homogeneity of s^*_m:

$$\Phi_{\tilde{m}}(x) \propto (-x)^{\beta_{\varepsilon,L}} \qquad \text{for } x \leq 0 \tag{16}$$

$$\Phi_{\tilde{\chi}}(x) \propto |x|^{-\gamma_{\varepsilon,L}} \tag{17}$$

c) g^*_c is analytic for all finite L and shows non-analyticities only in the TDL. To the best of our knowledge, no proof exists that s^*_m is analytic for finite systems. Depending on the values of $\beta_{\varepsilon,L}$ and $\gamma_{\varepsilon,L}$, Eqs. (16) and (17) imply the possibility for s^*_m to be either an analytic or a non-analytic function. This means that even for a completely analytic entropy, non-analyticities can occur for the microcanonical magnetization and susceptibility.

d) Note that it is indeed possible to explicitly construct a function $\tilde{s}^*_m$ which complies with the requirements (i)-(iii) stated at the end of the introduction. This explicit form is quite informative with respect to the "sudden" change of the exponents from their finite system values towards their values in the TDL, but has to be discussed elsewhere [6].

3 Conclusion

A MFSS theory has been developed in accordance with the demanded properties (i)-(iii) stated above. Amazingly, although the scaling laws (7) and (9) comprise identical degrees of homogeneity a_ε and a_m for the singular parts of the entropy of the finite <u>and</u> infinite system respectively, they nevertheless can account for power law behaviour with different exponents $\beta_{\varepsilon,L}$, $\gamma_{\varepsilon,L}$ for finite and $\beta_{\varepsilon,\infty}$, $\gamma_{\varepsilon,\infty}$ for infinite systems. As canonical potentials emerge from the microcanonical ones basically by means of a Laplace transform, it is to be expected that the smoothing properties of this integral transform cause MFSS to be applicable for smaller systems than CFSS.

References

1. In the case of an Ising system, the specific enthalpy ε is given by the interaction energy: $\varepsilon := -L^{-d} J \sum_{\langle i,j \rangle} S_i S_j$, i.e., ε does not contain the field dependent term $h \sum_i S_i$ of the Ising Hamiltonian.
2. M. Kastner, M. Promberger, A. Hüller, in "Computer Simulation Studies in Condensed Matter Physics XI", Eds. D.P. Landau, H.-B. Schüttler, Heidelberg (1998)
3. The zero-field magnetization and the susceptibility can be expressed in terms of ε, ε^* or $\tilde{\varepsilon}$. The corresponding functions are denoted by m, χ; m^*, χ^* or $\tilde{m}, \tilde{\chi}$.
4. C. DiCastro, G. Jona-Lasinio, in "Phase Transitions and Critical Phenomena" Vol. 6, Eds. C. Domb, M.S. Green, London (1976)
5. H.E. Stanley, "Introduction to Phase Transitions and Critical Phenomena", Appendix C, Oxford (1971)
6. M. Kastner, M. Promberger, A. Hüller, to be published
7. M.E. Fischer, M.N. Barber, Phys. Rev. Lett. **28**, 1516 (1972)

Molecular Dynamic Simulation of Directional Crystal Growth

B.V. Costa[1], P.Z. Coura[2], and O.N. Mesquita[1]

[1] Universidade Federal de Minas Gerais, Departamento de Física ICEx, C.P. 702, CEP 30123-970, Belo Horizonte - MG, Brazil
[2] Universidade Federal de Juiz de Fora, Departamento de Física

Abstract. We use molecular dynamic to simulate the directional growth of binary mixtures. our results compare very well with analytical and experimental results. This opens up the possibility to probe growth situations which are difficult to reach experimentally, being an important tool for further experimental and theoretical developments in the area of crystal growth.

1 Introduction

The rapid expansion of the use of high quality crystalline materials in optical and eletronic devices has strongly stimulated research, both theoretical and experimental, on dynamics of crystallization. Computer simulation has played an important role on the development and understanding of crystal growth. During growth the solid-fluid interface can display several interesting phenomena like segregation, dynamical instabilities and pattern formation [1].

A crystal can grow from the adjacent fluid by different mechanisms, depending on the structure of the interface (rough or smooth), material purity, growth rates, temperature gradients and so on. For a crystal to grow atoms or molecules must be transported from the fluid towards the interface with a non-zero sticking probability. besides that, the latent heat generated at the interface as well as the solute excess segregated must be carried away. These requirements can be met in a controlled way by puting the sample in an appropriated furnace submmited to a temperature gradient and pulling it with constant speed toward the colder region. results of such experiments have been compared with results of two dimensional models. Nijmeijer and Landau[2] reported molecular dynamic (MD) simulation on laser heated pedestal growth of fibers. Previous simulations consisted of kinetic models using Monte Carlo techniques and numerical solution of transport equations[3].

2 Simulation

In this work we use MD to simulate the solidification of a two component system of a solvent (particles of type a) and solute (particles of type b) interating via a Lennard-Jones (LJ) potential. By tuning the LJ parameters

Springer Proceedings in Physics, Volume 85
Computer Simulation Studies in Condensed-Matter Physics XII
Eds.: D.P. Landau, S.P. Lewis and H.-B. Schüttler

we set the structure of the interface (rough or smooth) and the segregation coefficient. We also simulate morphological instabilities, the planar inteface becoming cellular and eventually dendritic (Mullins-Sekerka instability.). The simulation is carried out with all particles interacting through the LJ potential

$$\phi_{i,j}(r_{i,j}) = \epsilon_{i,j}\left[(\frac{\sigma_{i,j}}{r_{i,j}})^{12} - (\frac{\sigma_{i,j}}{r_{i,j}})^{6}\right]. \tag{1}$$

The indices i and j stand for particles in the positions r_i and r_j respectively and $r_{i,j} = |r_i - r_j|$. There are three types of interactions, solvent-solvent, solvent-solute and solute-solute. In each case the LJ parameters are labeled as (a, a), (a, b) and (b, b). distance r, time t and temperature T are measured in units of $\sigma_{a,a}$, $\sigma_{a,a}(m_a/\epsilon_{a,a})^{\frac{1}{2}}$ and $k_B/\epsilon_{a,a}$. Initially we distribute $N = n_x \times n_z$ particles over the two dimensional volume $L_x \times L_z$. We assume periodic boundary conditions in the x direction. In the z direction we divide the system in to two distinct regions, a solid and a fluid one. In the solid region particles stand initially in their equilibrium position in a total of $n_x \times n_{0z}$ particles. on the fluid region we distribute the rest of the particles with the density chosen to be $\rho = 0.5\sigma_{a,a}^{-2}$, randomly distributed in a triangular lattice and slightly dislocated from their equilibrium position. We impose a temperature gradient along the z direction using a velocity renormalization approach. The system is divided in two regions, one cold which is set to $T = 0$ and the other to $T = T_h$ higher than the melting temperature T_m which is $T_m = 0.4$ in our units. we let the system evolve in time for N_e time steps of size δt until it reaches equilibrium. Once the equilibrium is reached we start pulling the system in the $+z$ direction, at a pulling velocity V_p. The $-z_{max}$ layer works as a particle source, maintaining a constant flow to the material. The basic experimental set up is shown in Fig. 1

In Fig. 2 we show an example of solute segregation during directional growth of the binary mixture: caprolactane as solvent and methyl-blue as solute. In the top part of Fig. 2 is shown an image of the crystal (left side) and melt (right side), with maximum concentration of methyl-blue at the melt side of the interface. From the gray level of the image we obtain the methyl-blue concentration profile across the sample, which in the melt, decays exponetially as a function of the distance from the interface (bottom part of Fig. 2). Fig. 3 shows the simulation results done in a system initially with 27×30 particles, solute concentration of 5%, pulling velocity $V_p = 0.004$, and LJ parameters given by: $\epsilon_{a,b} = 0.5$, $\epsilon_{b,b} = 0.1$, $\sigma_{a,b} = 1.0$, $\sigma_{b,b} = 1.0$ and $m_b = 1.0$.

After averaging over many runs to improve statistics one obtains the steady-state solute concentration profile represented as data points with error bars.

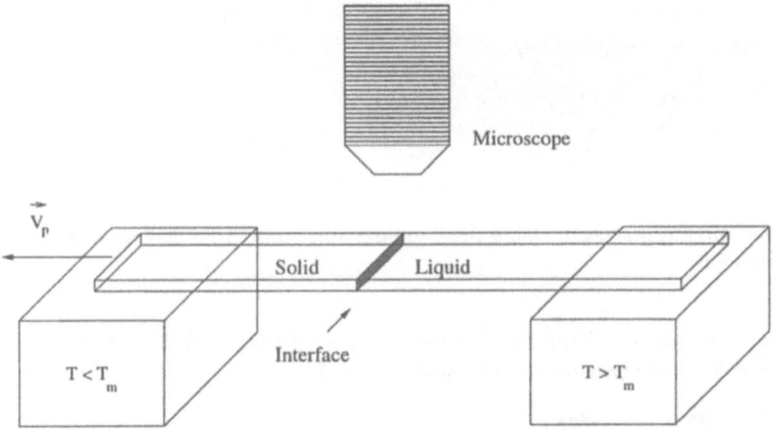

Fig. 1. Basic experimental set up for directional growth. The interface motion is visualized using an optical microscope. T_m is the melting temperature of the mixture and V_p is the pulling velocity

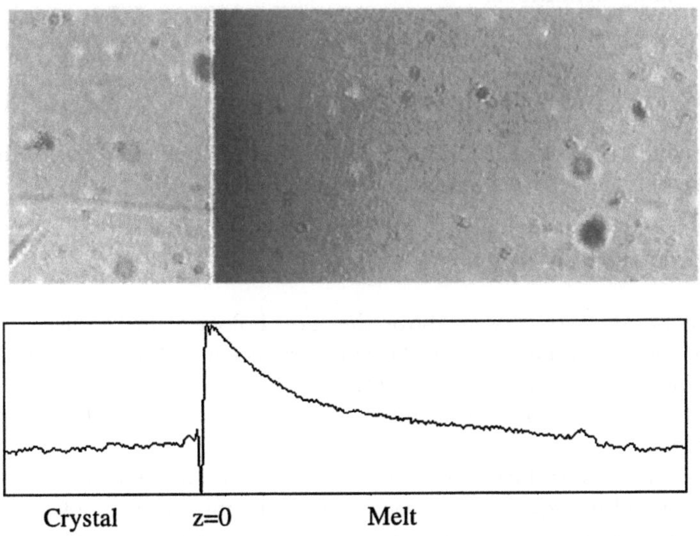

Fig. 2. Directional solidification of a binary mixture showing the solute segregation at the interface. The experimental result is for the mixture caprolactane (solvent) and methyl-blue (solute)

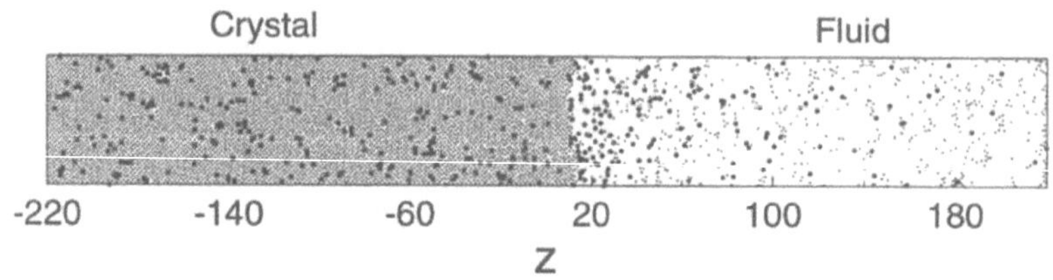

Fig. 3. Our simulation results using the Lennard-Jones potential describes in detail the experiment

In Fig. 4 we show a theoretical result obtained from the diffusion equation for the solute concentration in the fluid phase[4]

$$\frac{\partial c_f}{\partial t} = D\frac{\partial^2 c_f}{\partial z^2} + V_f\frac{\partial c_f}{\partial z} \tag{2}$$

where D is the solute diffusion coefficient in the fluid phase and V_f is the velocity of the system of reference. The diffusion equation is supplemented with the boundary conditions:

$$c_f = c_0 \qquad at \;\; z \to \infty\,, \tag{3a}$$

$$(1-K)V_f c_f = -D\frac{dc_f}{dz} \qquad at \;\; z = 0\,. \tag{3b}$$

The second condition is just the mass conservation at the interface and K is the ratio between the solute and solvent concentrations.

3 Cellular Instability

With the LJ parameters chosen above the morphological instabilities were inhibited during the cellular growth. We can stimulate instabilities by tunning the LJ parameters, particularly by slightly varying the equilibrium position between particles a and b, i.e., changing $\sigma_{a,b}$. Fig. 4 show the effect of varying this parameter.

For $\sigma_{a,b} = 1$ the interface is rough, becoming cellular and dendritic for $\sigma_{a,b} > 1$. As should be expected a lower $\sigma_{a,b}$ sets the interface smoother than larger values indicating that the system is very sensitive to geometric factores. (We have also simulated some experiments fixing the value of $\sigma_{a,b}$ and varying $\epsilon_{a,b}$ and $\epsilon_{b,b}$, the results did not show any qualitative change in their behavior. Raising $\sigma_{a,b}$ breaks the hexagonal structure o fthe solid. The energy involved in such distortion is so large that the interface segragate solute particles, bonds (a,a) and (b,b) are preferible than (a,b). Once particles of the type b concentrate at the interface its melting point is lowered and channels of b particles are formed inside the crystal. Fig. 5 Shows this effect.

We are now studying the stability [5] of those interfaces. We start with a smoth interface then, we introduce a perturbation at the interface. It has been

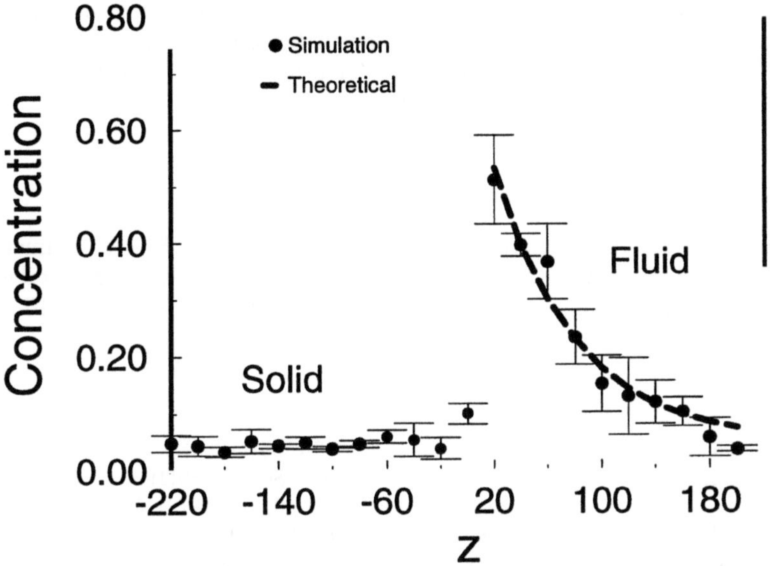

Fig. 4. The figure shows the concentration profile found in our simulation. It is to be compared to Fig. 2

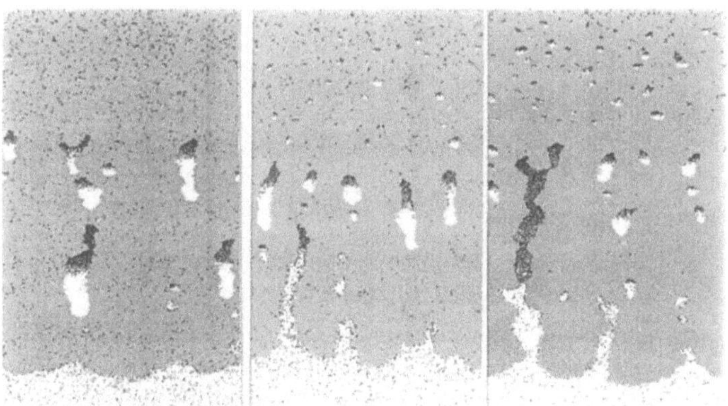

Fig. 5. Cellular instability originated by varying $\sigma_{a,b}$. From left to right $\sigma_{a,b} = 1.05, 1.10$ and 1.20

shown that near the bifurcation from planar to cellular the time evolution of the most instable Fourier mode of the perturbation can be described by a third order Landau-amplitude equation. This however is the subject of a new research and will be soon published elsewhere.

Acknowledgements

This work was partially supported by CNPq, FAPEMIG and FINEP. Numerical work was done at the CENAPAD-MG/CO and in the LINUX parallel cluster at the *Laboratório de Simulação* Departamento de Física - UFMG.

References

1. M.C.Cross and P.Hohenberg, Rev. Mod. Phys. **65**, 851(1993).
2. M.J.P.Nijmeijer and D.P.Landau, Comput. Mater. Sci. **1**, 389(1993).
3. J.D.Weeks and G.H.Gilmer, Adv. Chem. Phys. bf 40, 157(1979), D.P.Woodruff, The solid-liquid interface (Cambridge University Press, London, 1973).
4. P.Z.Coura, O.N.Mesquita and B.V.Costa, Phys. Rev B bf 59, 8345(1999).
5. J.M.A.Figueiredo, M.B.L.Santos, L.O.Ladeira and O.N.Mesquita, Phys. Rev. Lett. bf 71, 4397(1993).

Part V

Polymers and Surfactants

Escape Transition of a Grafted Polymer Chain

A. Milchev[1], V. Yamakov[1], and K. Binder[2]

[1] Institute for Physical Chemistry, Bulgarian Academy of Sciences, 1113 Sofia, Bulgaria
[2] Institut für Physik, Johannes-Gutenberg-Universität, Staudinger Weg 7, D-55099 Mainz, Germany

Abstract. The escape transition of a flexible polymer chain of chain length N, end-grafted at a hard wall and compressed by a piston of radius R in good solvent conditions, is studied by Monte Carlo simulation and by phenomenological arguments. In contrast to previous theories which have predicted a jump in the force f at a critical value H_t of the height H of the piston above the wall, we find that the transition (which is sharp only for $N \to \infty$) is characterized by a flat region of f in the $f - H$ isotherm, i. e. a jump in the height occurs at the transition from $H_{esc,t}$ to $H_{imp,t}$, with $(H_{imp,t} - H_{esc,t})/H_{esc,t} \approx 0.26$. At the transition the constant force f_t is predicted and observed to scale with R and N as $f_t \propto R^{\frac{1+\nu}{1-\nu}}/N^{\frac{2\nu}{1-\nu}}$ where the Flory exponent $\nu \approx 0.589$.

1 Introduction

Deformations of polymers at interfaces by applied forces occur in the context of colloid stabilization, biopolymers at cell membranes, interaction between polymers and AFM tips, etc. [1,2]. Due to their flexibility, the polymer chains may undergo conformational transitions induced by such forces. Much recent interest has found the prediction [3–6,9] that a polymer mushroom compressed by a cylindrical disk of radius R, that is much larger than the linear dimension R_0 of the undeformed polymer ($R_0 \propto lN^{\nu}$, a being the segment length and $\nu \approx 0.59$ [10]) but much smaller than its linear extension lN, undergoes an "escape transition" (Fig. 1). For $H \ll R_0$ but still exceeding some transition value H_t the mushroom gets deformed into a quasi-two-dimensional ($d = 2$) conformation "imprisoned" underneath the disk, until at a certain height of the disk above the grafting surface part of the chain "escapes" this "confinement", moving out into the free space around the piston. This conformational transition is purely entropic in nature and is driven by the entropy gain of the escaped part of the chain - cf. Fig. 1b. Theoretically it is expected[3–6] that the typical behavior of a first-order phase transition should be observed with a characteristic hysteresis in terms of the force, f, applied to the AFM-tip, and the height of the tip, H, over the surface. Of course, one should keep in mind that a sharp phase transition for single polymer chains may only occur for infinitely long chains $N \to \infty$ while for finite chain length the transition should be somewhat rounded due to finite size effects.

Springer Proceedings in Physics, Volume 85
Computer Simulation Studies in Condensed-Matter Physics XII
Eds.: D.P. Landau, S.P. Lewis and H.-B. Schüttler

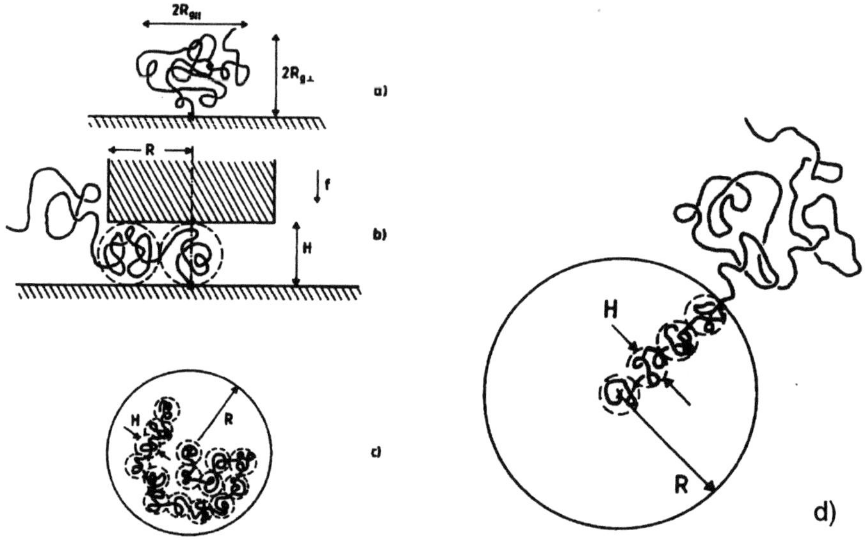

Fig. 1. a) Schematic side view of a "mushroom" (i. e., a chain grafted with one end at a flat repulsive wall). b) Compression of a mushroom by a piston of radius R which has its lower surface at a height H above the grafting surface. A force f is required to keep the piston at this height. The part of the chain under the piston forms a chain of n_t blobs of diameter H, while in the situation shown part of the chain is escaped. c) Top view of a compressed mushroom that is completely "imprisoned" under the piston. d) Top view of an "escaped" chain with a strongly stretched part

In this talk we present an attempt to test the theoretical predictions, which were mostly derived by scaling arguments. We use Monte Carlo simulations of a bead-spring coarse-grained model. This approach automatically includes statistical fluctuations which were neglected in the theoretical works[3–6], and therefore allows one to check to what extent the transition is rounded due to finite size effects.

In what follows we give a brief presentation of the basic theoretical arguments[3] in Section 2. The Monte Carlo model is sketched in Section 3 while the most important results from the simulations are given in Section 4. We conclude in Section 5 with a brief summary of our results.

2 Theoretical Background

Here we briefly repeat and expand the scaling treatment that was first presented by Subramanian et al.[3–6] for grafted polymer chains in the case of a good solvent. Considering a polymer mushroom under a disk at height $H \ll R_0$ but still above the transition height H_t, the mushroom gets deformed into a quasi-two-dimensional conformation with a lateral gyration radius (note that[10] $\nu = 3/4$ in $d = 2$), cf. (Fig.1c),

$$R_{g\parallel} \propto (H/l)^{1-\frac{3}{4\nu}} l N^{\frac{3}{4}}. \tag{1}$$

Eq.(1) derives from the concept that the chain can be divided into N' blobs with $N' = N/n_b$ where the number of monomers in a blob $n_b \propto (H/l)^{1/\nu}$ since[10,11] $H \propto l n_b^{\nu}$. Here and in what follows l denotes the average bond length between neighboring beads along the polymer. The free energy of the chain then is [3,10] $F_{imp} \propto k_B T N' \propto k_B T N (l/H)^{1/\nu}$.

However, when H gets even smaller, a transition to a partially escaped configuration of the chain may occur (Fig. 1b,d). Then only $n_t < N'$ blobs remain under the piston while the rest of the monomers are in one large blob outside. The tethered chain attains then a quasi-one-dimensional conformation with its part under the piston being pulled by the "escaped" chain fraction. This transition is driven by the gain in configurational entropy due to the part of the chain that has "escaped". Subramanian et al. [3] estimate the free energy of this state as

$$F_{esc} \approx F_0 + F_{str} \propto k_B T \left[n_t + \left(\frac{R}{H}\right)^4 \frac{1}{n_t^3} \right], \tag{2}$$

where $F_0 = k_B T n_t$ would be the free energy of a two-dimensional chain of blobs having its natural size $\left(R_0 \propto H n_t^{3/4}\right)$, while F_{str} is the free energy cost of stretching this chain *strongly* up to the radius R, namely[12] $F_{str} = k_B T (R/R_0)^4$. Minimizing F_{esc} with respect to n_t

$$\frac{\partial F_{esc}}{\partial n_t} = 0 \Rightarrow n_t = 3^{\frac{1}{4}} \frac{R}{H}, \tag{3}$$

then yields the simple result[3] $F_{esc} \propto k_B T (R/H)$. Subramanian et al. [3–6] then locate a first order transition at $H = H_t$ between these states by equating $F_{esc} = F_{imp}$, which yields

$$H_t/a \propto (lN/R)^{\nu/(1-\nu)}, \tag{4}$$

and is accompanied by a jump in the force $f = -(\partial F/\partial H)_T$ at the transition.

From Eqs.(1), (4) one can estimate the size of the imprisoned chain at the transition height:

$$R_{g\parallel}^t \propto l N^{\frac{3}{4}} \left(\frac{R}{lN}\right)^{\frac{3-4\nu}{4(1-\nu)}} . \tag{5}$$

Thus at the transition the linear dimension $R_{g\parallel}$ undergoes a first-order transition from $R_{g\parallel}^t \ll R$ to $R_{g\parallel}^t \propto R$.

Since we consider here two rather distinct states, qualitatively described by Fig. 1c and by Fig. 1d, which coexist at $H = H_t$ with each other, it also is of interest to estimate the *free energy barrier* between them. This free energy barrier is described by a state where the imprisoned chain is sufficiently stretched such that the parallel component of its end-to-end distance $R_{g\|} = R$, since then the chain end can move out underneath the piston and lower the free energy by the configurational entropy gained by the monomers that have escaped there. The free energy can be estimated from Eq.(2) using $n_t^t = N' = N/n_b \approx N\,(H_t/l)^{-1/\nu}$, n_t^t being the number of blobs of the "imprisoned" chain at the transition. This yields

$$\frac{F_{barrier}}{k_B T} \approx N' + \left(\frac{R}{R_{g\|}}\right)^4 . \tag{6}$$

and inserting now Eq.(4) yields (both terms in Eq.(6) are then of the same order)

$$\frac{F_{barrier}}{k_B T} \propto N \left(\frac{R}{lN}\right)^{\frac{1}{1-\nu}} . \tag{7}$$

We are now in a position to estimate qualitatively the region in the space of parameters $\{R, N\}$ where this escape transition occurs (of course, for finite N this transition always will be rounded due to statistical fluctuations). One obvious condition is that $H_t/l > 1$: we cannot squeeze the chain more than until it takes a fully two-dimensional configuration, $R_{g\|} \propto lN^{3/4}$. From Eq.(4) we then conclude that the maximum radius $R_{\max}$ of the piston in Figs.1, for which an escape transition can occur at fixed finite height H_t, is

$$R_{\max}/l \propto N \quad , \quad N \to \infty \quad . \tag{8}$$

In this extreme case H_t/a stays finite as $N \to \infty$, and $F_{barrier}/k_BT$ is proportional to N, so that a pronounced hysteresis of the escape transition can be expected.

The opposite limit occurs when $F_{barrier}/k_BT \approx 1$. Then the transition is completely washed out and an absolutely smooth variation of the chain radius $R_{g\|}$ with H is expected. From Eq.(7) we immediately recognize that this limit occurs for

$$R_{\min}/l \propto N^{\nu} \quad , \quad N \to \infty \quad . \tag{9}$$

When $R = R_{\max}$ we see from Eq.(5) that then $R^t_{g\|}/R \propto N^{-1/4} \to 0$ as $N \to \infty$ while for $R = R_{\min}$ we have $R^t_{g\|}/R \approx const$. Hence we conclude that escape transitions occur in the shaded region of the plane $(1/R, 1/N)$ in Fig.2.

Here we note that there is an important difference between this conclusion and the results of our investigations[7]. The reason for this discrepancy is in fact very simple: H is an *extensive* thermodynamic variable, while the force f is the conjugate *intensive* variable. We have an analogy with the liquid-gas transition (H corresponds to the volume V, f to the pressure p) where it is

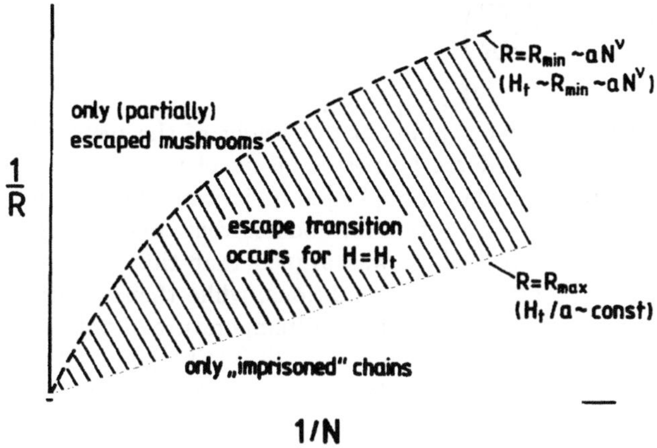

Fig. 2. Phase diagram in the plane of variables - inverse chain length $1/N$ and inverse piston radius $1/R$ - showing the region where escape transitions can occur. Note that a sharp transition occurs only for $1/N \to 0$ and $R_{\min} \ll R \leq R_{\max}$. For $R \to R_{\min}$ the escape transition gets progressively rounded, since $F_{barrier}/k_BT$ decreases from order N for $R = R_{\max}$ to order unity for $R = R_{\min}$. Actually our treatment implies that there are two transition heights, $H_{esc,t}$ and $H_{imp,t}$ cf. Eqs.(13a,13b), rather than a single one

the Gibbs potential $G(T,p) = f(T,V)+pV$, which is equal in both phases at the transition pressure $p_{coex}(T)$, while for V one has a jump from V_{liq} to V_{gas}. We have to use a similar Legendre transform here, $G(T,f) = F(T,H) + fH$. Writing $F_{esc} = c'k_BTR/H$ and $F_{imp} = ck_BTN(l/H)^{1/\nu}$, we obtain[8]

$$G_{esc}(T,f) = F_{esc}(T,H) + fH = 2\sqrt{c'k_BTf} \tag{10}$$

$$G_{imp}(T,f) = \nu^{1/(1+\nu)}(1+\frac{1}{\nu})(ck_BTNl^{1/\nu})^{\nu/(1+\nu)}f^{1/(1+\nu)}, \tag{11}$$

Now the transition is located from $G_{esc}(T,f_t) = G_{imp}(T,f_t)$ at

$$\frac{lf_t}{k_BT} = \left[\frac{\nu^{\frac{1}{1+\nu}}}{2}(1+\frac{1}{\nu})\right]^{2\frac{1+\nu}{\nu-1}} \frac{(c'\frac{R}{l})^{\frac{1+\nu}{1-\nu}}}{cN^{\frac{2\nu}{1-\nu}}}, \tag{12}$$

and the associated values of the height, $H_{esc,t}$ and $H_{imp,t}$ which limit the region where the escaped state ($H > H_{esc,t}$) or the imprisoned state ($H < H_{imp,t}$) represent the thermal equilibrium are

$$\frac{H_{esc,t}}{l} = \left(\frac{1+\nu}{2}\right)^{\frac{1+\nu}{1-\nu}} \nu^{\frac{\nu}{\nu-1}} \left(\frac{cNl}{c'R}\right)^{\nu/(1-\nu)}, \tag{13a}$$

$$\frac{H_{imp,t}}{l} = \left(\frac{1+\nu}{2}\right)^{\frac{2\nu}{1-\nu}} \nu^{\frac{\nu}{\nu-1}} \left(\frac{cNl}{c'R}\right)^{\nu/(1-\nu)} . \tag{13b}$$

Using $\nu \approx 0.59$, one finds $(H_{imp,t} - H_{esc,t})/H_{esc,t} \approx 0.26$.

Thus our considerations predict a jump in the height of the piston (at constant force f) rather than in the force f, exerted on the piston at constant height H, as suggested by Subramanian et al.[3–6] in their original treatment. But even the very existence of this transition for good solvent conditions has been called into question recently [10]. It was argued that this transition may only occur for bad and Theta solvent conditions, while then a critical temperature T_c is reached, analogous to the critical temperature of polymers in dilute solution in bad solvents, so that for $T > T_c$ (in the good solvent regime) this transition is completely eliminated by statistical fluctuations.

3 The Model

We have studied this problem by Monte Carlo simulation of an off-lattice bead spring model which has been previously proven valuable for chains both in the bulk and near surfaces [13–15]. In this model coarse-grained effective bonds are represented by springs described by the finitely extensible nonlinear elastic ($FENE$) potential,

$$U_{FENE} = -\frac{K}{2} R^2 \ln\left[1 - (l - lo)^2/R^2\right], \qquad l_{\min} < l < l_{\max}, \tag{14}$$

with l being the bond length, $R = l_{\max} - l_0$, and l_0, $l_{\max}$, $l_{\min} = 2l_0 - l_{\max}$ being the equilibrium value of the bond length and its maximum or minimum extension, respectively. As before [13–15], parameters were chosen such that $l_{\max} = 1$ as the unit of length and $l_0 = 0.7$, $K = 40$ (in units of k_BT).

The non-bonded interactions between beads are described by the Morse potential,

$$U_M(r) = \varepsilon\{\exp[-2\alpha(r - r_{\min})] - 2\exp\left[-\alpha(r - r_{\min})\right]\}, \tag{15}$$

where r is the distance between beads, and the parameters are chosen as $r_{\min} = 0.8, \varepsilon = 1$, and $\alpha = 24$. Due to the large value of α, $U_M(r)$ decays to zero very rapidly for $r > r_{\min}$, and is completely negligible for $r > 1$ already. These choices of parameters allow the use of a very fast link-cell algorithm. Choosing a temperature $k_BT = 1.0$, we are in the good solvent regime of the model, well above its θ-temperature ($k_B\theta \approx 0.62$ [13]).

The impenetrable surface plane is at $z = 0$, so the beads can only be located in the half space $(x, y, z > 0)$, the grafted end being at the coordinate origin - Fig. 3. The center of the hard cylindrical piston is always at the line $x = y = 0$, exactly above the grafted end, and the height of the piston H above the surface and its radius R are the parameters which are varied in our simulations as well as the chain length (choosing $N = 128$, 256, 512, 768, 1024

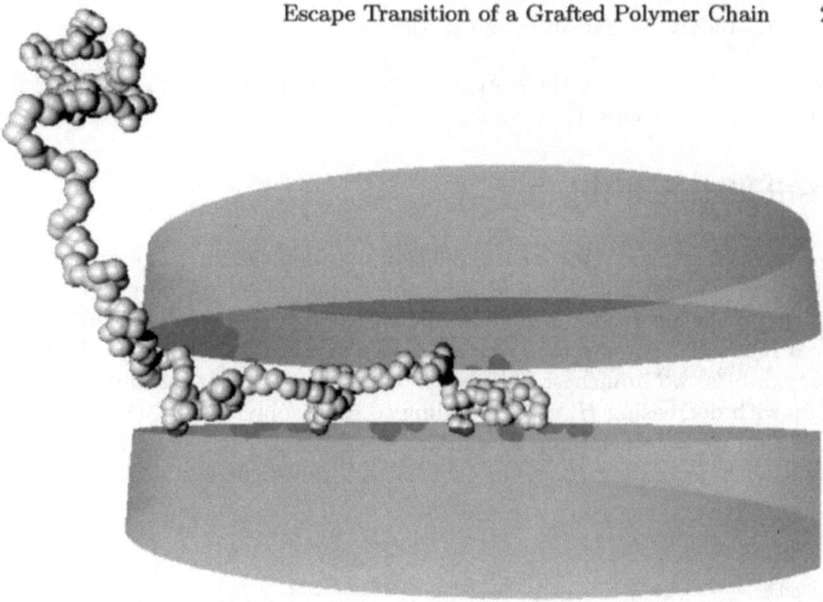

Fig. 3. Snapshot of a partially "escaped" polymer chain with $N = 256$ monomers, grafted to a flat surface and compressed by a piston with radius $R = 20$ at height $H = 5$

and 1280, respectively). Clearly, the choice of the piston center exactly above the grafted end is somewhat special, but arguments have been presented [6] that the escape transition persists also for off-centered piston positions. The most symmetric case studied here should thus be taken as a generic example only.

As a simulation technique, we choose a combination of enhanced configurational bias methods[7,16] and simple random hopping moves[13] for generating well-equilibrated initial states. A combination of 8 random hopping moves (Monte Carlo steps) per monomer and a two- and three-dimensional pivot move[17] has turned to be especially efficient for the Monte Carlo averaging. We also employed an *expanded* ensemble technique[18] with trial transitions between neighboring states (i.e. heights H_m) with *preweighting* factors η_m. The grand partition function Ω is given by the sum over the canonical partition functions $Z_m(T, H_m)$ of, say, M different states (heights)

$$\Omega = \sum_{0}^{M} Z_m(N, H_m) \exp(\eta_m) \tag{16}$$

and the transition from state m to n is accepted with probability

$$P(n \to m) = \min\{1, \exp[-(E_m - E_n + \eta_m - \eta_n)]\} \tag{17}$$

with weights $\eta_m = -\int_0^m f(H)dH + \eta_0$. Technical details about these (rather involved) algorithms, which allow the study of much longer chain lengths rather than when one uses the simple random hopping algorithm only [13,14], are presented elsewhere [7].

4 Simulational Results

Fig. 4 presents typical data for the relation between the force exerted by the polymer on the piston (measured from the z-component of the virial tensor [13,15]) and the height, for a number of choices of R and N (we consider only choices $R \gg l_0 N^{\nu}$, of course). We can recognize clearly that the f vs. H relation exhibits two branches: for large H there is a branch that increases less steeply with decreasing H, corresponding to imprisoned chains, then there is a transition region resembling a "van der Waals loop", and again a a steeper increase with decreasing H sets in - this latter branch corresponds to escaped

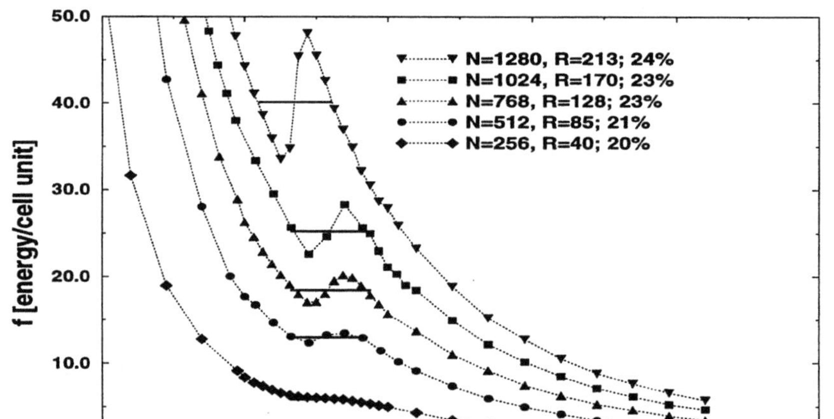

Fig. 4. Normalized force $l_{\max} f/k_B T$ plotted vs. distance H, for several choices of N and R, as indicated in the figure. The ratio $(H_{imp,t} - H_{esc,t})/H_{esc,t}$ is given in brackets in the legend for each set. $H_{esc,t}$ and $H_{imp,t}$ are measured as the left and right side of the transition region, marked by the horizontal section, drawn according to the Maxwell rule, for each set

chains. Actually the loops in the transition region in Fig. 4 have metastable and unstable parts, in analogy to the loops at the liquid-gas transition. In fact, increasing the simulation time substantially the extent of the metastable branches gets somewhat smaller. Thus the equilibrium behavior is estimated by a Maxwell construction (horizontal straight lines in Fig. 4). For the longer

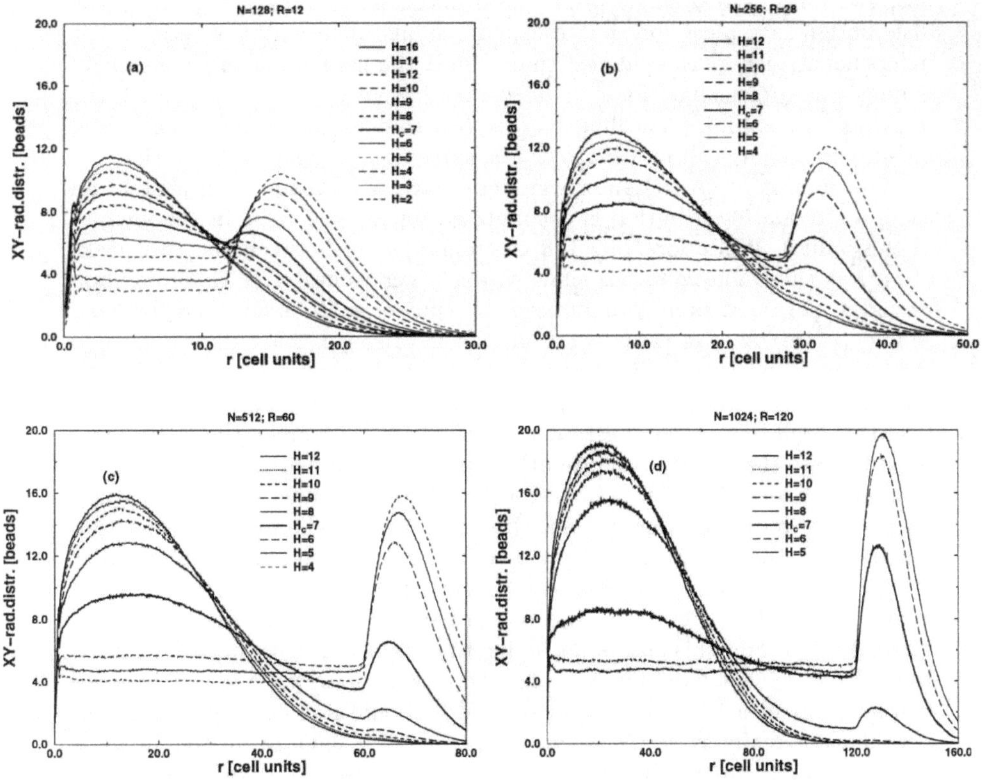

Fig. 5. Radial density distribution function $\rho(r)$ of the monomers in a compressed mushroom for different choices of R and N. Several values of H are included, as shown in the figure

chains a check of the ratio $(H_{imp,t} - H_{esc,t})/H_{esc,t} \approx 26\%$ in a good agreement with Eq.(12). It is also evident that the coexistence region becomes pronounced for sufficiently long chains $N \geq 256$ whereas for shorter chains the phase transition is "washed out" by fluctuations.

The assignment of different states belonging to the two branches can be made very directly either by inspection of the chain configurations or by recording the radial monomer density distribution $\rho(r)$, where $r = \sqrt{x^2 + y^2}$ and x, y, z are the coordinates of a monomer (Fig. 5). One sees that in the "imprisoned" state of a chain the distribution rises fast to a maximum at a distance $r_{\max} \ll R$, and then it decreases steadily until at $r \approx 5r_{\max} \div 8r_{\max}$ it is essentially zero. Decreasing H first leads to a slight shift of $r_{\max}$ to larger values, but then the shape of $\rho(r)$ changes qualitatively: it becomes much broader, and a side peak at $r > R$ develops. These distributions correspond to the flat part of the isotherm in Fig. 4. Finally $\rho(r < R)$ is completely flat, and for $r > R$ a single pronounced maximum occurs: this form of the distribution is typical for the "escaped" chains. The region where $\rho(r)$ is independent of r gives evidence that indeed the configuration of the chain is quasi-one-dimensional for $r < R$, as sketched in Fig.1c. As expected, the position $r_{\max}$ exhibits a scaling $r_{\max} \propto N^{\nu}$, that is, the weakly compressed mushroom resides still in a more or less three-dimensional configuration.

The choices of (N, R) shown in Fig.5 have been made such that in all cases a bimodal distribution is encountered for $H_c = 7$ (and in the vicinity of this value). While for $N = 128$ the minimum in between the two peaks of $\rho(r)$ at this value is rather shallow, for larger values of N this minimum becomes more and more pronounced. Of course, as H is changed for fixed R and (finite) N, the crossover from the imprisoned chain distribution to the escaped one is a gradual transition, as it should be for any finite N, but the point is that the transition does become sharper as $N \to \infty$. If we take this limit at a fixed constant ratio of R/lN (of course, this ratio must be much less than unity because otherwise the chain can never escape), we ultimately expect for $\rho(r)$ a behavior with two delta-functions, if we study ρ as a function of the rescaled coordinate $\tilde{r} = r/R$. The one delta-function representing the imprisoned chain is at the origin (since a radius of order lN^{ν} for $N \to \infty$ is negligible on a scale lN), while the other delta function is at $\tilde{r} = 1$ in that limit (the size of the escaped "blob" in Fig.1 is also of the order lN^{ν} and this size is negligible in comparison with both lN and R when R/lN is kept fixed for $N \to \infty$). Obviously, even for $N = 1024$ we are still far from such a limiting behavior, and it should also be noted that these data are not really at $R/lN = const$, since $R/N = 0.09375$ in Fig. 5a while $R/N = 0.1172$ in Fig. 5d. From our results it becomes immediately clear that it is very important to study the rounding of the escape transition due to finite chain length; this subject will be taken up in a forthcoming paper. From the analysis of Fig.5 one can also show that the escape transition is not characterized by a single value H_t, but rather a kind of two-phase coexistence region, $H_{esc,t} \leq H \leq H_{imp,t}$ - Eqs.(13a,13b). Thus one could estimate the contributions of the two states in the region of heights H where a bimodal distribution is observed in terms of the heights h_1, h_2 of the peaks above the intervening minimum. Of course, in principle at first-order transition one should obtain the "weights" of the two phases from suitably normalized areas underneath the two peaks, but since the distinction of what belongs to one peak and what belongs to the other peak is somewhat arbitrary, we did not try to do this and restricted ourselves to a qualitative analysis with this heuristic criterion[7]. Indeed, a plot of the relative weight, $h_1/(h_1+h_2)$ versus H_c where we define H_c from the condition $h_1 = h_2$ shows[7] that all curves intersect in a single point $h_1/(h_1+h_2) = 0.5, H/H_c = 1$ by construction. The increase of $h_1/(h_1 + h_2)$ for $N \to \infty$ is very nearly linear between 0 and 1 with a slope of $(H_{inp,t} + H_{esc,t})/2/(H_{inp,t} - H_{esc,t}) \approx 4.35$, matching closely the prediction of Eqs.(13a,13b).

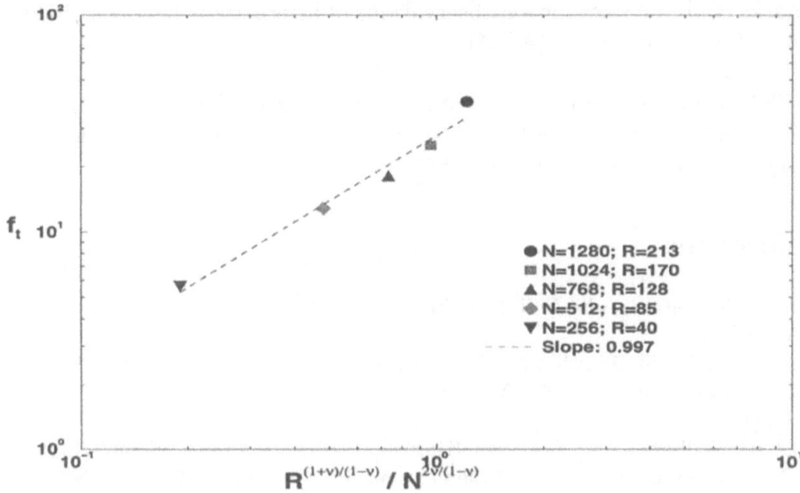

Fig. 6. Variation of the equilibrium force f_t at the escape transition with the ratio $R^{\frac{1+\nu}{1-\nu}}/N^{\frac{2\nu}{1-\nu}}$, cf. Eq.(12). The values of f_t were determined from the isotherms on Fig.4

Finally we would like to verify the prediction of our analytical considerations, Eq.(12) that the equilibrium value of the strength f_t, applied to the piston, scales with R and N in a specific way - Fig. 6. Although f_t could be determined graphically from Fig.4 with moderate accuracy only, it is evident from Fig.6 that the prediction of Eq.(12) is confirmed. In any case, we can refute the claim of Jimenez and Rajagopalan[9] that in the excluded volume case an escape transition does not exist at all - we think their failure to observe the escape transition in the good solvent regime is simply because they focused attention on a single and fairly short ($N = 100$) chain length.

5 Summary

In summary, we have presented the first evidence from simulations that when compressing a polymer mushroom by a cylindrical piston under good solvent conditions one will encounter an escape transition, as predicted recently [3–6]. Although for finite N inevitably a rounding of the transition takes place, clear signals of the transition can both be recorded from the force vs. height isotherm (Fig. 4) and from the bimodal character of the radial monomer density distribution (Fig. 5). However, in thermal equilibrium there is no jump of the force at the transition height, instead there is a regime of *constant force* for $H_{esc,t} < H < H_{imp,t}$. Of course, it is conceivable to follow the loops containing metastable and unstable pure escaped (imprisoned) states inside this "forbidden" region, as our simulation shows. We expect that these remarks could be relevant for observations of interactions between chains and atomic force microscope (AFM) tips.

***

This work was supported in part by the Deutsche Forschungsgemeinschaft (DFG) under grant N° 436BUL113/92 and by the Bulgarian National Science Foundation under grant X-640/1997.

References

1. P. G. de Gennes, Adv. Coll. Interface Sci. **27** (1987) 187; Adv. Polymer Sci. **138**, 90(1999).
2. A. Halperin, M. Tirrell, and T. P. Lodge, Adv. Polymer Sci. **100**, 33(1992).
3. G. Subramanian, D. R. Williams, and P. A. Pincus, Europhys. Lett. **29**, 285(1995).
4. D. R. Williams, and F. C. MacKintosh, J. Phys. II France, **5**,1407(1995).
5. G. Subramanian, D. R. Williams, and P. A. Pincus, Macromolecules, **29**, 4045(1996).
6. M. C. Guffond., D. R. Williams, and E. M. Sevick, Langmuir, **13**, 5691(1997).
7. A. MIlchev, V. Yamakov and K. Binder, Phys. Chem. Chem. Phys. (in press).
8. For the sake of completeness, we allow constants of order unity, c, c' here.
9. J. Jimenez, and R. Rajagopalan, Langmuir **14**, 2598(1998); Europ. Phys. J. B **5**, 237(1998).
10. P. G. de Gennes, *Scaling Concepts in Polymer Physics,* Cornell University Press, Ithaca 1979.
11. By the symbol $\propto$ we mean that prefactors of order unity are suppressed.
12. P. Pincus, Macromolecules, **9**, 387(1976).
13. A. milchev, W. Paul, and K. Binder, J. Chem. Phys. **99**, 4786(1993); A. MIlchev and K. Binder, Macromol. Theory Simul. **3** 915(1994); V. Yamakov, A. MIlchev and K. Binder, J. Phys. II France, **7**, 1123(1997).
14. , A. MIlchev and K. Binder, Macromolecules, **29**, 343(1996); J. Phys. II, France, **6**,21(1996); Europ. Phys. J. B **3**, 477(1998).
15. A. Milchev, R. Pandey, and K. Binder, Macromolecules, **30**, 1194(1997).
16. D. Frenkel and B. Smit, *Understanding Molecular Simulation: From Algorithms to Applications*, Academic Press, New York (1996) pp. 292-300
17. N. Madras, and A. D. Sokal, J. Stat. Phys. **50**, 109(1988).
18. N.B.Wilding and M. Müller, J.Chem.Phys.**101**, 4324-4330 (1994)

Monte Carlo Simulations of Model Nonionic Surfactants

A.P. Chatterjee and A.Z. Panagiotopoulos

Institute for Physical Science & Technology,
University of Maryland,
College Park, Maryland 20742, USA

Abstract. The focus of this work is on aggregation and micellization in a lattice model for nonionic surfactants. Formation of micellar aggregates as a function of temperature and surfactant chemical potential was studied by histogram reweighting grand canonical Monte Carlo simulations. Two different sets of site-site interactions were utilized. The critical micellar concentration (CMC) was determined from the equation of state obtained from simulations in small systems. Results were obtained for the micellar size distributions. The introduction of effective repulsive site-site interactions between nearest-neighbour head-head and head-tail groups, in addition to the attractive tail-tail segmental interactions considered in earlier investigations, was shown to cause a significant decrease in the CMC. It was also found to attenuate the dependence of micellar size on surfactant chemical potential. The observed dependence of the CMC on temperature for the present model disagrees with experimental results for the n-alkyl m-ethoxyl series of surfactants in water. This is attributable to the failure of the model to take into account the (entropic) hydrophobic effect important for amphiphilic aggregation in aqueous solutions.

1 Introduction

The formation of self-assembled aggregates in aqueous solutions of amphiphiles has been investigated using experimental and theoretical methods [1–4], and, more recently, computer simulation studies [5–9]. As the surfactant concentration is increased at a fixed temperature, various macroscopic properties of the solution, e.g. the electrical conductivity or osmotic pressure, show abrupt changes in their dependence on surfactant concentration [3]. These changes are associated with the formation of mesoscopically ordered amphiphilic aggregates, which may be of spherical, cylindrical, or lamellar architecture. The overall surfactant concentration at which such aggregates first form is defined as the critical micellar concentration (CMC). The CMC generally depends on temperature, surfactant architecture and the presence of additional components in solution. For overall surfactant concentrations above the CMC, the concentration of free (monomeric) amphiphiles in the bulk is observed to be relatively constant, and the added surfactant leads to an increase in the number or size of the micellar aggregates. Formation of mesoscopically ordered microphases has been studied extensively for the case of relatively high molecular weight diblock copolymer systems [10]. Such

Springer Proceedings in Physics, Volume 85
Computer Simulation Studies in Condensed-Matter Physics XII
Eds.: D.P. Landau, S.P. Lewis and H.-B. Schüttler

systems may be regarded as amphiphiles with a large degree of polymerization and exhibit very rich phase behavior at temperatures low enough that the incompatibility between the segments in the individual blocks leads to microphase separation [12].

In recent years, several computer simulation based studies have examined the aggregation process for amphiphiles of lower molecular weight (compared to block copolymeric systems) for surfactant molecules having both linear as well as "gemini" architectures. The pioneering studies in this context were initiated by R.G. Larson [13], who examined equilibria between surfactant microphases at relatively low temperatures and high surfactant volume fractions using lattice Monte Carlo simulations performed in the canonical (constant NVT) ensemble for model molecules with short-ranged nearest-neighbor interactions. More recently, the greater flexibility of the grand canonical ensemble (constant μVT conditions) together with histogram reweighting techniques have been employed in the low (near CMC) surfactant concentration regime to investigate the onset of micellization and the dependence of the CMC on temperature [9]. These studies have focussed primarily on model amphiphilic molecules of symmetric architecture, and a single fixed choice of the microscopic interactions describing head-head, head-tail, and tail-tail interactions has been employed. The present study explores a situation in which the model surfactant molecule may be of asymmetric structure, reflecting the relative volumes of head and tail group segments of real experimental hydrophobic and hydrophilic moieties, and also investigates in a preliminary manner the effects of using a different choice of microscopic interaction strengths.

2 Models and Methods

The model studied consists of linear amphiphile molecules, each occupying a set of connected sites on a simple cubic lattice. All vacant sites, i.e., those that are not occupied by either hydrophilic head (H) or hydrophobic tail (T) sites, are considered to be occupied by structureless solvent particles. The specific amphiphilic molecule we investigate in this study has the structure H_9T_6; if we identify each lattice site as having a volume of approximately $\approx 60\mathring{A}^3$, this structure has head and tail group volumes which correspond roughly to those of the hydrophilic and hydrophobic portions of the nonionic n-dodecyl octaoxyethylene glycol ether surfactant $C_{12}E_8$, the aggregation of which has been extensively investigated experimentally under aqueous conditions [14].

Two sets of site-site interaction sets were studied; in both cases, the only interactions assumed to be present (besides excluded volume) were between nearest-neighbor head-head (HH), head-tail (HT), and tail-tail (TT) contacts, where the set of nearest-neighbours of a given site refers to the 26 locations given by the relative vectors (1,0,0), (1,1,0), (1,1,1), and their symmetry-equivalent analogues. Periodic boundary conditions are employed in all directions. The first interaction set employed in this work (denoted I1) corresponds to the choice used in earlier work on related models for other, shorter, amphiphile chains, and specifies the pairwise additive thermal interactions as being [0,0,-1.3333] for the HH, HT, and TT interactions, respectively. Interaction set I1 thus includes only attractive tail-tail interactions; the only head-head and head-tail interactions present are purely of the excluded volume type, as only a single mer is permitted to occupy a given lattice site. The specific value for the TT attractive interaction used in this work was chosen to maintain equality of the thermal chi-parameter on a per-amphiphile basis with earlier work on surfactants with the H_4T_4 architecture, which had employed the interaction set [0,0,-2]. This form (I1) of interaction set was used in previous simulation studies of amphiphilic aggregation, as well as in the initial work by R. Larson which focussed on ordered mesophases which occur at high (liquid-like) amphiphilic volume fractions. The second interaction set we study (denoted I2) has HH, HT, and TT nearest-neighbor interactions given by [+2,+1,-2], respectively; as for the I1 set, positive/negative interaction energies correspond to repulsive/attractive interactions. Interaction set I2 thus includes repulsive HH and HT interactions over and above the excluded volume restriction. The temperature scale is in both cases set by the energy units employed in the interaction sets.

We performed Monte Carlo (MC) simulations of the one-component amphiphile system in the grand ensemble. The move types employed include insertion and removal of chains, chain conformational renewal by reptation moves, and also translational moves of connected amphiphile "clusters". The insertion/removal of chains was performed using the athermal configuration-bias technique to permit chain insertions without overlapping with pre-existing occupied sites. Amphiphile "clusters" are identified as sets of chains such that every included chain shares in at least one nearest-neighbor TT interchain contact with another (distinct) chain contained in the same cluster. We typically employed 50 % transfer (chain insertion or removal), 49.9 % reptation-type, and 0.1 % cluster moves in the course of a run; the moves resulting in displacement of a connected cluster are computationally relatively expensive, and hence attempted infrequently. Systems were equilibrated at each temperature and chemical potential over a period of $10 - 30 \times 10^6$ MC steps, followed by collection of data over a subsequent period of $30 - 50 \times 10^6$ steps. Simulation runs were performed at a series of different temperatures and chemical potential (μ) for each interaction set; histograms of the frequencies of occurrence of total amphiphile numbers and total system energy were combined and re-weighted to obtain estimates of the grand partition function, and hence the compressibility factor, as functions of the temperature and imposed chemical potential:

$$\frac{PV}{k_B T} = \ln \Xi = \ln(\sum_{N,E} e^{(N\mu - E)/k_B T}), \tag{1}$$

where Ξ is the grand canonical partition function. As has been pointed out previously [9], significant hysteresis effects exist at low enough temperatures, with the state of aggregation (micellized or dispersed) and number of micelles observed being roughly constant over an entire run.

From the combined histograms, we determined the volume fraction occupied by amphiphiles and the compressibility factor as functions of the chemical potential. The same calculation allows determination of the compressibility factor as a function of amphiphile volume fraction at a given temperature. Simulation runs were performed for each temperature at a low enough chemical potential that the ideal gas behavior was recovered explicitly for low enough amphiphile volume fractions. At sufficiently low temperatures (as discussed in greater detail subsequently), the ideal gas behavior breaks down as the amphiphile volume fraction is raised, giving way to a weaker but still demonstrably linear growth in the pressure with equilibrium amphiphile volume fraction. This weaker growth in pressure with amphiphile volume fraction is due to the incipient aggregation into micellar clusters, which leads to a net decrease in the number of independent kinetic entities in the simulation box relative to the scenario in which all the amphiphiles are dispersed and unclustered. At high enough volume fractions of amphiphile, this second linear regime for the growth of the pressure eventually breaks down. The amphiphile volume fraction at the crossover between ideal gas behavior and the weaker, but still linear, pressure growth regime is determined by the intersection of piecewise linear fits to the pressure-volume fraction curves in both regimes; this concentration is employed as an operational estimate of the temperature-dependent CMC for our model system. Examples of the dependence of the compressibility factor on amphiphile volume fraction and temperature are shown in Fig. 1.

As discussed in more detail in the following section, our model exhibits a clearly identifiable CMC-like concentration only if the temperature is below a certain critical upper bound. As the temperature rises, the slope of the pressure-volume fraction curve in the secondary linear regime becomes larger, and eventually it becomes difficult to clearly distinguish this from the ideal gas regime. The present work regards an acceptably clear demarcation of the CMC as being possible so long as the slope of the pressure in this secondary concentration regime is no larger than 0.1. As mentioned in the Introduction, it is to be noted that precise location of the CMC depends on the property whose dependence on surfactant concentration is used to characterize the aggregation process (for polydisperse aggregates). Use of a metric other than the osmotic pressure may thus lead to a slight change in the CMC location even within the present model, but its dependence on temperature and choice of interaction set is not expected to be affected.

The following section presents results for the temperature dependence of the observed CMC for the I1 and I2 interaction sets, and for the enthalpic

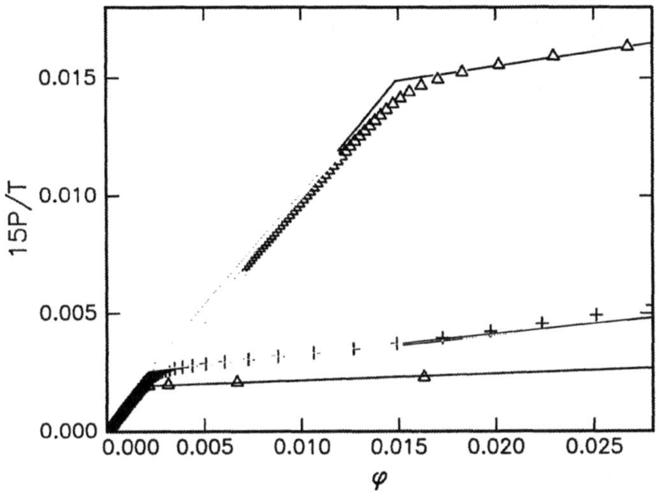

Fig. 1. The quantity $15P/k_BT$ as a function of the amphiphile volume fraction, ϕ, for H_9T_6 chains. The triangles show results for the I1 interaction set at $T = 6$ (upper) and $T = 5$ (lower); the crosses show results for the I2 interaction set at $T = 6.5$. The straight line segments at low amphiphile volume fraction correspond to ideal gas behaviour, while at higher volume fractions, the best linear fit to the secondary regime where P increases linearly with ϕ is shown

and entropic contributions to the free energy of micellization. Results for the micellar sizes, size distributions and radii of gyration are also presented.

3 Results

3.1 Critical Micelle Concentration (CMC)

GCMC simulation results for the CMC at various temperatures for the I1 interaction set, in which the amphiphiles were placed on a 30x30x30 sized simulation cell, are presented in Table 1; corresponding results for the I2 interaction set, also obtained for a 30x30x30 sized cell, are shown in Table 2.

The results in Tables 1 and 2 show that: (i) for *both* interaction sets I1 and I2, the CMC concentration increases with increasing temperature, and (ii) that the absolute values of ϕ_{cmc} are significantly reduced for the I2 interaction set as compared to values obtained using the interaction set I1. The observed increase in ϕ_{cmc} with increasing temperature in our model is qualitatively consistent with the experimental trend found for non-ionic surfactants of the n-alkyl m-oxyethylene form in weakly polar solvents, e.g., formaldehyde, and for α- monoolein in nonpolar solvents like benzene or cyclohexane [15]. However, this is inconsistent and of opposite nature to what

Table 1. Amphiphile volume fractions (ϕ_{cmc}) and chemical potentials (μ_{cmc}) at the CMC for interaction set I1 as a function of temperature T; also shown is the slope s of the compressibility factor as a function of amphiphile volume fraction in the secondary linear regime with amphiphile concentrations just above the CMC

T	ϕ_{cmc}	μ_{cmc}	s
5	0.00193	-44.2	0.028
5.5	0.00626	-42.1	0.049
5.6	0.00758	-41.7	0.063
5.7	0.00906	-41.4	0.075
5.8	0.0108	-41	0.087
5.9	0.0127	-40.7	0.104
6	0.0149	-40.4	0.123
6.5	0.0312	-38.8	0.207

Table 2. Amphiphile volume fractions ϕ_{cmc} and chemical potentials μ_{cmc} at the CMC for interaction set I2 as a function of temperature T; also shown is the slope s of the compressibility factor as a function of amphiphile volume fraction in the secondary linear regime with amphiphile concentrations just above the CMC

T	ϕ_{cmc}	μ_{cmc}	s
5.5	0.00030	-50.8	0.035
6	0.00098	-48.2	0.057
6.5	0.0026	-45.7	0.086
6.6	0.0032	-45.2	0.093
6.7	0.0037	-44.8	0.107
6.8	0.0043	-44.3	0.123
6.9	0.0050	-43.9	0.134
7	0.0056	-43.6	0.177
7.5	0.0118	-40.9	0.256

is found experimentally for surfactant molecules of the n-alkyl m-ethoxyl family in aqueous solutions, for which the CMC is known to decrease with increasing temperature [14]. The reason for this qualitative inconsistency with the experimentally observed behavior of non-ionic surfactants of the C_nE_m family in water becomes clearer upon performing straight-line fits to the logarithm of ϕ_{cmc} versus $1/T$, the slopes and intercepts of these fits yielding the enthalpy and entropy of micellization:

$$\ln(\phi_{cmc}) = \frac{\delta G_{mic}}{k_B T} = \frac{\delta H_{mic}}{k_B T} - \frac{\delta S_{mic}}{k_B}, \qquad (2)$$

where $\delta G_{mic}, \delta H_{mic}$, and δS_{mic} are the standard Gibbs free energy, enthalpy, and entropy of micellization per amphiphile molecule, respectively. (The volume fraction of the pure solvent/"water" has been equated to unity in applying the result due to Molyneux et. al. [16]). The logarithm of ϕ_{cmc} is shown as a function of $1/T$ for both interaction sets I1 and I2 in Fig. (2); the fitted straight lines yield: $\delta H_{mic} \approx -60.1, \delta S_{mic} \approx -5.83 k_B$ for the I1 interaction set, and $\delta H_{mic} \approx -75.2, \delta S_{mic} \approx -5.60 k_B$ for interaction set I2. In both cases, both the entropy and enthalpy of micellization is found to be negative, i.e., the formation of aggregates is an energetically favorable, but entropically hindered process. Based on the operational criterion that a well-defined CMC is to be identified only under conditions such that the slope of the compressibility factor as a function of amphiphile volume fraction does not exceed 0.1 in the secondary linear regime which follows the ideal-gas like behavior, there exists an "upper critical micellization temperature" at about $T \approx 5.9 \pm 0.1$ for the I1 and $T \approx 6.7 \pm 0.1$ for the I2 interaction sets.

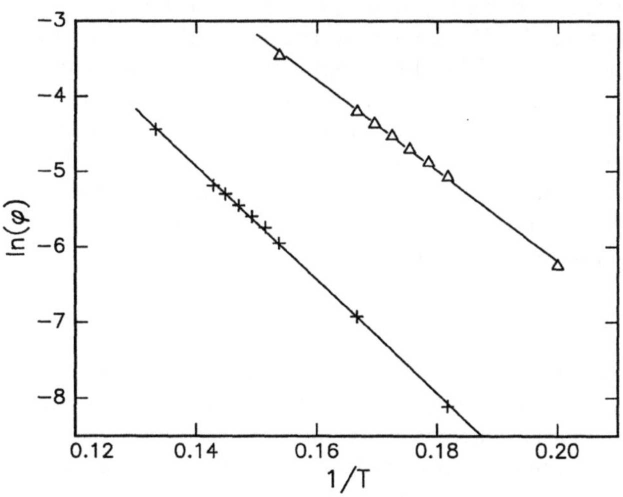

Fig. 2. $\ln(\phi_{cmc})$ is plotted as a function of $1/T$ for the I1 (triangles) and I2 (crosses) interaction sets; the lines are linear fits using the values of δH_{mic} and δS_{mic} discussed in Section III

The absolute values of the CMC found for the model H_9T_6 molecule for both interaction sets are somewhat higher than the experimentally determined CMC for the $C_{12}E_8$ amphiphile in water. For example, at room temperature (298 K), $C_{12}E_8$ in water has a CMC of about 7.1×10^{-5}M; if we use the mapping that each unit cell on our lattice corresponds to a volume of roughly 60Å^3, this concentration corresponds to an amphiphile volume fraction of $\phi_{cmc} \approx 3.8 \times 10^{-5}$. Thus, at the lowest temperatures studied presently, the CMC values for our H_9T_6 model exceed the experimental room temperature values for $C_{12}E_8$ by factors of about 8 and 50 for the I2 and I1 interaction sets, respectively. Inclusion of repulsive head-head and head-tail interactions significantly lowers the absolute value of the model CMC in the range of temperature we have been able to study, and brings it closer to the experimental range of values at room temperature, but it is still too large by a significant factor.

Our present results for the enthalpic contribution to the free energy of micellization for the I1 case may be compared with the findings of earlier work on amphiphiles of the H_3T_3 and H_4T_4 architectures, which had employed an I1-like interaction set specified by [0,0,-2] for the HH, HT, and TT interactions, respectively. Taking together the earlier and present results for δH_{mic} for H_3T_3, H_4T_4, and H_9T_6 surfactant molecules, the quantity $\delta H_{mic}/\epsilon_{TT}$ varies in the ratio of approximately 3:4:5.7 for this set of architectures. Thus, for interaction sets of the I1 type, the enthalpy of micellization is approximately proportional to the number of interacting tail groups per surfactant, and always favors aggregation. The slightly more positive value for $\delta H_{mic}/|\epsilon_{TT}|$ for the I2 with respect to the I1 interaction sets for H_9T_6 (-37.6 versus -45.2) reflects the head-head and head-tail repulsive interactions which are augmented when the head group segments of aggregated chains are forced to inhabit the corona region surrounding the micellar core. Examination of earlier and present results (for the I1 case) for the entropy of micellization δS_{mic} per amphiphile reveal that, for the H_3T_3, H_4T_4, and H_9T_6 set of amphiphiles, the findings are consistent with a linear dependence of $\delta S_{mic}/k_B$ on the number of tail groups N_T *alone*:

$$\frac{\delta S_{mic}}{k_B} = -1.18 - 0.77 N_T. \tag{3}$$

Note that this quantity is (at least for the H_9T_6 molecule of the present study) *not* dramatically altered on going from the I1 to the I2 interaction set, suggesting that the major entropic contribution in the micellization process stems from the re-organization of tail group segments on going from the dispersed to micellar core regions. These remarks should be regarded as merely suggestive, as they are based on results for only four combinations of amphiphile architecture and interaction set. In all cases, the entropic contribution favors amphiphilic dispersion rather than aggregation.

For the experimentally relevant n-alkyl m-ethoxyl surfactants, precisely the *opposite* situation is known to be true [14]; thus, δH_{mic} and δS_{mic} are *both* positive, indicating that the formation of aggregates is an entropically rather than energetically-driven process. The underlying source for this phenomenon is the "hydrophobic effect", first studied in detail by Frank and Evans [17] in the context of the water solubility of simple, non-ionic molecules. Our present model does not do justice to the structure of water as a hydrogen-bonded, networked fluid; in particular, the potential for formation of "icebergs" in the water domains surrounding dispersed amphiphile molecules, leading to a locally more ordered and hence lower entropy structure, is not captured by a simple model in which water molecules are represented as sites interacting via purely isotropic nearest-neighbor interactions. The fact that we find δH_{mic} to be negative for our model for both interaction sets is consistent with the situation that in either case, the formation of aggregates must be accomplished based on purely energetic considerations. In recent years, a number of models for water have been proposed which incorporate these aspects relevant to a faithful description of the aggregation phenomenon in aqueous environment [18]. The use of such models for the solvent/"water" together with our present treatment of the amphiphilic molecules may be expected to yield results in closer qualitative accord with experiment, and shall be pursued in future work on this subject.

3.2 Aggregate Size

Results for the oligomer and micellar volume fractions and micellar aggregation numbers, $< M >$ are shown in Tables 3 and 4 for the I1 and I2 interaction sets, respectively. In each case, the cluster size distribution (frequency of occurence of clusters of a given size) was obtained from a GCMC run equilibrated at the relevant values of T and μ, and the peak found at higher values of M was fitted to a Gaussian distribution. The standard deviation of the Gaussian fitted curves, W, are also shown on Tables 3 and 4. Also shown in Tables 3 and 4 are the mean values of the three principal radii of gyration as calculated from snapshots of the bead coordinates, averaged over all clusters containing more than 10 amphiphile molecules. The location of the center of mass of the micelle is calculated by assigning equal weights to head and tail group segments. The error estimates in the tables represent standard deviations of results from four independent runs.

The results in Tables 3 and 4 show that, as estimated from the radii of gyration, the micellar dimensions are roughly similar for both interaction sets studied; however, the mean aggregation number is significantly lower for the I2 interaction set. For both interaction sets, a systematic decrease in the micelle size is found with increasing temperature. This is consistent with our observation that, above an "upper critical micellization temperature," we do not find formation of stable micellar aggregates for either the I1 or I2 interaction sets, as the aggregation process is energetically driven for the present model.

Table 3. Mean micellar aggregation numbers $< M >$, averaged values of the principal radii of gyration, and dispersed amphiphile (oligomeric clusters) and micellar volume fractions ϕ_{olig} and ϕ_{mic} for interaction set I1. Numbers in parentheses indicate error estimates (one standard deviation) in the last digit, where available

T	μ	ϕ_{olig}	ϕ_{mic}	$< M >$	W	$R_g^{(1)}$	$R_g^{(2)}$	$R_g^{(3)}$
5.5	-41.5	0.0058(0)	0.031(1)	56(2)	9.3	5.17(8)	4.98(7)	4.69(7)
5.5	-41.1	0.0061(2)	0.036(3)	66(5)	9.7	5.4(1)	5.1(1)	4.9(1)
5.5	-41	0.0064(0)	0.037(1)	67(2)	10.2	5.36(6)	5.17(5)	4.90(6)
5.5	-40.9	0.0053(2)	0.073(3)	66(2)	10.0	5.34(8)	5.1(1)	4.89(9)
5.5	-40.6	0.0040(2)	0.114(6)	70(4)	10.4	5.45(7)	5.25(7)	5.00(7)
6	-39.6	0.019(1)	0.009(6)	46(4)	9.2	4.8(2)	4.7(2)	4.3(3)
6	-39	0.015(2)	0.05(2)	51(2)	9.7	5.13(6)	4.91(8)	4.60(7)
6.5	-37	0.034(2)	0.042(6)	37(1)	11.0	4.9(1)	4.65(9)	4.23(4)

Table 4. Mean micellar aggregation numbers $< M >$, average values of the principal radii of gyration, and dispersed amphiphile (oligomeric clusters) and micellar volume fractions ϕ_{olig} and ϕ_{mic} for interaction set I2. Numbers in parenthese indicate error estimates (one standard deviation) in the last digit where available

T	μ	ϕ_{olig}	ϕ_{mic}	$< M >$	W	$R_g^{(1)}$	$R_g^{(2)}$	$R_g^{(3)}$
5	-51.4	0.000027(1)	0.067(2)	42(2)	4.1	5.48(6)	5.31(8)	5.05(8)
5.5	-49.4	0.00026(1)	0.0184(7)	33(2)	4.7	5.23(4)	5.05(4)	4.80(4)
5.5	-49.3	0.00025(1)	0.028(3)	49(6)	5.9	5.8(2)	5.6(2)	5.4(2)
5.5	-49.2	0.00022(1)	0.030(2)	54(4)	3.5	5.91(8)	5.8(1)	5.55(8)
5.5	-48.9	0.00019(1)	0.043(3)	38(2)	3.8	5.42(9)	5.3(1)	5.03(8)
5.5	-46.5	0.00012(2)	0.082(6)	50(2)	4.3	5.7(1)	5.5(1)	5.3(1)
6.5	-43.3	0.0019(1)	0.038(3)	35(3)	5.4	5.3(1)	5.1(1)	4.9(1)
6.5	-43	0.0013(2)	0.056(2)	35(1)	5.4	5.31(4)	5.11(4)	4.86(3)
6.5	-42	0.0013(1)	0.063(2)	38(1)	5.0	5.45(5)	5.26(4)	5.01(4)
7	-40.5	0.0046(5)	0.036(5)	31(2)	5.0	5.2(1)	4.9(1)	4.7(1)
7	-38.5	0.0028(8)	0.069(8)	36(1)	6.0	5.35(7)	5.14(6)	4.90(7)
7.5	-38	0.0083(7)	0.040(4)	29(1)	5.9	5.08(3)	4.84(4)	4.53(5)
7.5	-37.5	0.008(2)	0.047(9)	29(1)	5.2	5.11(7)	4.88(7)	4.59(8)
7.5	-35.5	0.0051(9)	0.073(4)	34(1)	5.4	5.28(6)	5.05(7)	4.76(9)

4 Discussion

The thermodynamics of amphiphile aggregation into approximately spherical micelles has been investigated for a linear, asymmetric, model surfactant molecule on a lattice for two different sets of microscopic interaction parameters. Results for the CMC, and the enthalpy and entropy of micellization, have been compared to previous simulation results for shorter-chain, symmetric surfactant molecules. Issues relating to the micellar size (in terms of both mean aggregation number and radii of gyration) and structure (quantified by density profiles and orientational order parameters for the molecular bond vectors), together with a comparison to predictions of single chain mean-field theory, shall be addressed in a future work.

The primary result of the present work has been to show that the enthalpy and entropy of micellization are approximately linear in the number of interacting groups per molecule, for both symmetric and asymmetric architectures. In addition, we find that alteration of the microscopic pairwise interaction strengths can significantly impact the CMC concentration; this may be of great relevance to future studies designed to more closely approximate the situation in real aqueous surfactant assemblies. However, the qualitative dependence of CMC on temperature was not altered for the two interaction sets considered here, and remains at variance with the experimental trend known for the alkyl-ethoxyl family of amphiphiles in water. As discussed previously, we believe this inconsistency is a consequence of the present treatment of water as a structureless, single-site vacancy, which neglects the complex orientation and local density-dependence of the hydrogen bonding network characteristic of water at room temperatures. Consideration of more realistic lattice models for water, which incorporate aspects of the hydrophobic effect, is currently under way, and is expected to improve agreement with at least the qualitative features observed experimentally when used in conjunction with the present models for the surfactant-only system.

References

1. Tanford, C. (1980) The hydrophobic Effect. Wiley, New York.
2. Schick, M. J. (1987) Nonionic Surfactants; Physical Chemistry. Marcel Dekker, New York.
3. Rosen, M.J. (1989) Surfactants and Interfacial Phenomena. 2nd Ed., John Wiley, New York.
4. Israelachvili, J. (1992) Intermolecular and Surface Forces. 2nd Ed. Academic Press, London.
5. Von Gottberg, F. K., Smith, K. A., Hatton, T. A., (1997) Stochastic Dynamics Simulation of Surfactant Self-assembly. J. Chem. Phys. **106**, 9850-7.
6. Mackie, A. D., Panagiotopoulos, A. Z., Szleifer, I. (1997) Aggregation Behavior of a Lattice Model for Amphiphiles. Langmuir **13**, 5022-31.

7. Xing, L., Mattice, W. L. (1997) Strong Solubilization of Small Molecules by Triblock-Copolymer Micelles in Selective Solvent. Macromolecules **30**, 1711-1717.
8. Viduna, D., Milchev, A., Binder, K. (1998) Monte Carlo Simulation of Micelle Formation in Block Copolymer Solutions. Macromolecular Theory Simul. **7**, 649-58.
9. Floriano, M. A., Caponetti, E., Panagiotopoulos, A. Z. (1999) Micellization in Model Surfactant Systems. Langmuir, in press.
10. Almdal, K., Bates, F. S., Mortenson, K. (1992) Order, disorder, and Fluctuation Effects in an Asymmetric Poly (ethylene-propylene)-poly (ethylethylene) Diblock Copolymer. J. Chem. Phys. **96**, 9122-32.
11. Ferrenberg, A. M., Swendsen, R. H. (1988) New Monte Carlo Technique for Studying Phase Transitions. Phys. Rev. Lett. **61**, 2635-8.
12. Leibler, L. (1980) Theory of Microphase Separation in Block Copolymers. Macromolecules **13**, 1602-17.
13. Larson, R. G., Scriven, L. E., Davis, H. T. (1985) Monte Carlo simulation of amphiphile-oil-water systems. J. Chem. Phys. **83**, 2411-20.
14. Meguro, K., Takasawa, Y., Kawahashi, N., Tabata, Y., Ueno, M. (1981) Micellar Properties of a Series of Octaethyleneglycol-n-alkyl Ethers with Homogeneous Ethylene Oxide Chains and their Temperature Dependence. J. Coll. Interface Sci. **83**, 50-56.
15. Kon-no, K., Jin-no, T., Kitahara, A. (1974) Solubility, Critical Aggregating or Micellar Concentration, and Aggregate Formation of Nonionic Surfactants in Nonaqueous Solutions. J. Coll. Interface Sci. **49**, 383-89.
16. Molyneux, P., Rhodes, C. T., Swarbrick, J. (1965) Thermodynamics of Micellization of N-Alkyl Betaines. Trans. Faraday Soc. **61**, 1043-52.
17. Frank, H. S., Evans, M. W. (1945) Free Volume and Entropy in Condensed Systems - III. J. Chem. Phys. **13**, 507-32.
18. Roberts, C. J., Panagiotopoulos, A. Z., Debenedetti, P. G. (1996) Liquid-Liquid Immiscibility in Pure Fluids: Polyamorphism in Simulations of a Network-Forming Fluid. Phys. Rev. Lett. **77**, 4386-89; Roberts, C. J., Debenedetti, P. G. (1996) Polyamorphism and density anomalies in network-forming fluids: Zeroth- and first-order approximations. J. Chem. Phys. **105**, 658-72, and references therein.

A New Model to Simulate the Growth of Branched Polymers

H. Rêgo[1], L. Lucena[1], and G. M. Viswanathan[2]

[1] Departamento de Física Teórica e Experimental, Universidade Federal do Rio Grande do Norte, Natal-RN, Brazil
[2] Departamento de Física, Universidade Federal de Alagoas, Maceió-Alagoas, Brazil

Abstract. We propose a model to simulate branched polymers in disordered media, based on a model introduced by Lucena et al. [Phys. Rev. Lett. *72*, 230 (1994)]. In this new model we maintain the number of active tips fixed in the time, instead of growing as a power-law like in the original model. We compare our results with another growth mechanism based on the self-organized criticality (SOC) concept.

1 The Branched Polymer Growth Model

In past years the question of how polymerizaton occurs has became an important topic in experimental and theoretical studies [2–4]. It is well known that linear polymers in dilute solution can be modeled by self-avoiding walk (SAW) chains, where the SAW ensemble consists of all configurations of non-intersecting random walks [2,3]. In recent years it was found that large SAWs can be generated by kinetic growth model (KGM) in which, at each step, a random walker can move to a neighbor site that has not been visited before [5].

Some years ago a Branched Polymer Growth Model (BPGM) has been proposed by Lucena *et.al.* [1] to describe the growing process of branched polymers. By computational simulations they found a very rich phenomenology in this model (phase-transitions, non-linear behavior, avalanches and more recently self-organized criticality). The model itself is very simple. It contains the basic features of KGM and two more ingredients: a bifurcation probability b and a concentration of quenched impurities c.

In the BPGM, a branched polymer is generated from an initial seed in a d-dimensional lattice. In each step of the growing process, a bifurcation can occur with probability b. Additionaly, a fraction c of blocked sites is not available for the growing branched polymer (See Fig. 1). Numerical simulations in a square lattice show that there exists a critical line $b_c(c)$, where the correlation length diverges, which separates the infinite from the finite growth regime. Bunde *et. al.* [6,7] have identified these critical polymers as belonging to the same universality class of percolation.

Springer Proceedings in Physics, Volume 85
Computer Simulation Studies in Condensed-Matter Physics XII
Eds.: D.P. Landau, S.P. Lewis and H.-B. Schüttler

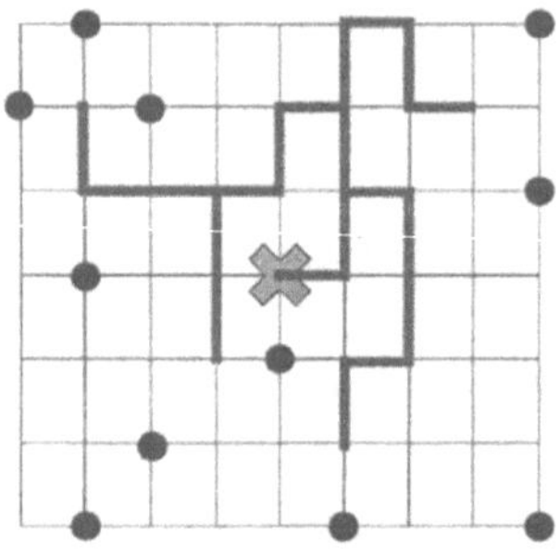

Fig. 1. Schematic representation of the Branched Polymer Growth Model (BPGM) in a square lattice. In the center of the lattice, the x marks the origin, where the branched polymer (wide line) start his growth process. The black dots are obstacles, representing the quenched impurities

2 Self-Organized Criticality in the Growth of Branched Polymers

More recently, Andrade Jr. *et. al.* [8] proposed a dynamical rule for the BPGM which regulates the flux of monomers available for agregation. Under this condition a self-organization process takes place and automaticaly leads the system to the criticality. This method was sucessfully used in the BPGM to obtain the critical line for the finite-infinite transition. Lucena *et. al* also used this method to determine the critical line that separates the regime in which the growing surface generated by the BPGM is rough, to the regime in which this surface is faceted. They called this a *faceted-to-rough transition* [8,9].

In this SOC like model the control rule is introduced to guarantee the infinite growth of the branched polymer, keeping the number of active sites N above zero. As a consequence, the control rule mantains the branching probability oscillating between the finite and infinite regimes, around the critical curve.

This is a powerfull method to calculate the critical line, not only because it is very simple, but also because it is very fast. One can determine a critical point averaging in time a single sample, and the accuracy will depend only on the running time.

In order to get a good accuracy it is necessary to simulate the system for long time intervals. If the control rule is applied in the parameter b, the system takes some time to feel the action caused by the feedback mechanism. When we apply the mechanism to the parameter c, the feedback response will appear in a more fast way.

In the other hand, sometimes this procedure is not very practical, specially if we are working in 3 dimensions. The fluctuations in the parameter used in the control rule are very strong in the tri-dimensional case, even when this parameter is the concentration of impurities. It means that we have to simulate the growing polymer in a huge lattice in order to get good accuracy for these critical points, and this is not always possible.

3 The Fixed Number of Tips Model (FNTM)

In the present paper, we propose a new model to simulate the growth of branched polymers that we believe can avoid this difficulty. Our model keeps the main ideas of the SOC model, but it is more simple than the former one.

We initialize our system with a finite number of active sites N, that we call active tips. For each following step, all active tips will randomly occupy an empty neighbor site. Eventually an active tip can be blocked by himself, due the self-avoiding effect, or by impurities forbiding his growth. We say that this tip dies. Everytime that this happens, we randomly choose one of the other active sites and imediatelly we introduce a bifurcation. As a direct consequence, the number of active tips will be rigorously constant in time. Hence we can guarantee that the branched polymer will grow indefinitely. Also, there will be no risk that the number of tips will grow exponentially. The polymer will be infinite, but the system does not penetrate in the compact region of the phase diagram. It will be always near the critical line.

Differently from the SOC model, the answer of the system in FNTM is instantaneous. Everytime an active tip is blocked, immediatly a new active tip sprouts from a different active site. The bifurcation probability b_{eff} is determined simply dividing the total number of births of active tips, B, by the total number of occupied sites M. If we think in terms of a real polymer, taking all sites with the same mass, we çan call M the total mass of the branched polymer.

By this way, it is not necessary simulate huge systems with large running times to get a good accuracy. To calculate the critical points, we can run during a small time several samples of a small system and take the average. The computational effort necessary to grow a cluster is lower, because we do not have to perform the numerical operations required by the control rule.

4 Results

From the computer simulations we can observe that the density distributions of sites occupied by the branched polymer generated by these two methods are approximately the same. Fig. 2 shows the comparison between the statistical density distributions of the two models. We note that for small radius, they are quite different, because we initialize the system with different number of active sites. For greater values of r, they have very similar distributions.

Therefore, as we can see also in the Fig. 2, we cannot say that the branched polymers generated by these two models are fractals. The mass M of the systems do not vary with r as a power-law.

Another remark is that our model cannot be reproduced in the laboratory, differently from the SOC model, in which Andrade Jr. *et.al.* suggested an experiment [8]. In this sense FNTM is a toy model.

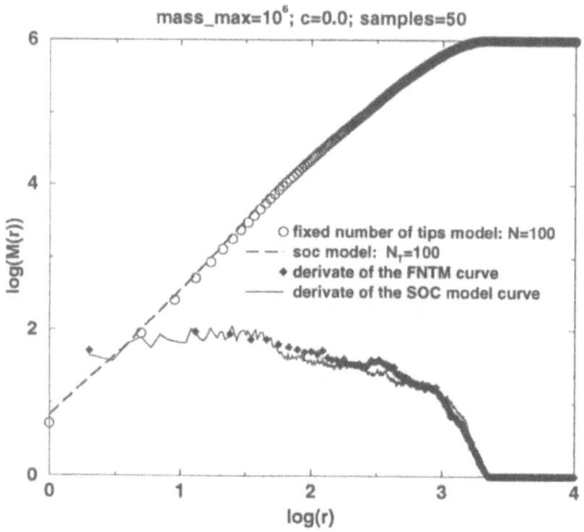

Fig. 2. Comparison between the statistical density distribution of monomers of polymers generated by our model and the self-organized model. In this case there is no impurities ($c = 0.0$). We also show the derivatives of the two curves with respect to $log(r)$. We maintain the number of active tips in our model N equal to 100 and we keep the average of active tips in the SOC model N_T around the same value

Similarly to the oscillations of the SOC model, b_{eff} will have a scattered distribution of values around the critical line, for different samples of the system. We take the average over the samples to calculate the critical line for the finite-infinite transition in a bi-dimensional system (See Fig. 3). Comparing this result with the obtained by the SOC method, we see that they have practically the same behavior.

5 Conclusion

The model proposed in this work is a toy model that presents some computational advantages compared with the previous ones.

In spite of its simplicity, it allows easy numerical calculation of the critical curve for the finite-infinite transition in $d = 2$ using small systems with small running time. This should be very usefull to determine the phase diagram of this transition, in a cubic lattice. We can also apply the same approach to find critical lines for other phase transitions even in different kind of lattices, like triangular, honeycomb, etc.

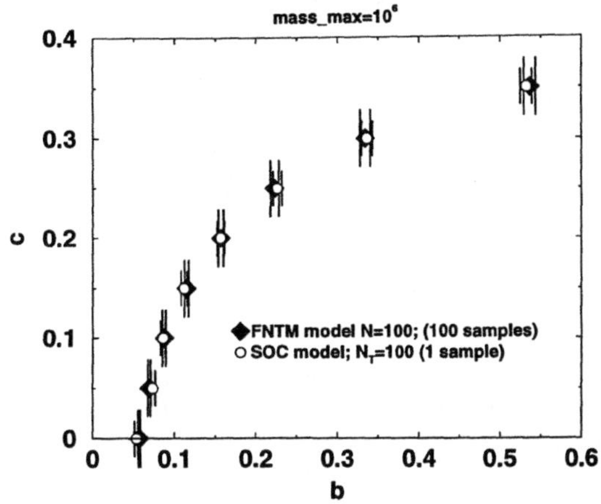

Fig. 3. The curves represents the critical line for the finite-infinite transition obtained by using our model (fixed number of tips model) and the SOC model. Using the last, we can determine the critical line with only one sample. In the FNTM we can get the same result using smaller systems and more samples. The taller parallel segments represents the error bars for the FNTM

References

1. Lucena, L. S. (1994) Branched Polymer Growth Model. Phys. Rev. Lett **72**, 230
2. Flory, P. J. (1953) Principles of Polymer Chemistry. (Cornell University Press, Ithaca)
3. de Gennes, P. G. (1979) Scaling Concepts in Polymer. Physics (Cornell University Press, Ithaca)
4. Doi, M., Edwards, S. F. (1986) The Theory of Polymer. Dynamics (Oxford University Press, Oxford)
5. Majid, I., Jan, N., Coniglio, A., Stanley, H. E. (1984) Kinetic Growth Walk. Phys. Rev. Lett **52** 1257
6. Bunde, A., Havlin, S., Porto, M. (1995) Are Branched Polymers in the Universality Class of Percolation? Phys. Rev. Lett. **74** 2714
7. Porto, M., Shehter, A., Bunde, A., Havlin, S. (1996) Branched Polymers in the presence of impurities. Phys. Rev. E **54** 1742
8. Andrade Jr, J. S., Lucena, L. S., Alencar, A. M., Freitas, J. E. (1997) Self-Organization in Growth of Branched Polymers. Physica A **238** 163-171
9. Lucena, L. S., da Silva, L. R., Roux, S. (1999) Faceted Rough Transition in BPGM. Physica A **266** 81-85

Molecular Dynamics Simulations of the Adsorption of Single Chains on Surfaces

A. Michel[1,2], S. Kreitmeier[2], and B. Wunderlich[1]

[1] University of Tennessee, Department of Chemistry, Knoxville TN 37996-1600, and the Chemistry and Analytical Sciences Division, ORNL, Oak Ridge TN 37831-6197, USA

[2] Universität Regensburg, Institut für Experimentelle und Angewandte Physik, Universitätsstr. 31, 93040 Regensburg, Germany

Abstract. Simulations of the adsorption behavior of single polymer chains on surfaces by means of isothermal molecular dynamics have been done. For the chain we are using a united atom model with stretching (two particle), bending (three particle) and dihedral (four particle) as well as a nonbonded interaction between nonadjacent atoms. The surface is given by a Lenard-Jones potential which represents in a first attempt a plain attractive surface. The shape of the experimentally generated and investigated single chains are compared to simulated conformations. Cristallisation effects in two dimensions have been found.

1 Introduction

Recent experiments on electrospraying of highly diluted solutions have shown that generating and determining the shape and size of isolated single polymer chains on a surface is possible [1]. Charged droplets consisting of solvent and polymer chains can be produced by this electrospray method. Drifting towards a counter electrode, the solvent evaporates and the charge density increases, promoting Rayleigh fission into smaller droplets. These particles hitting the surface sometimes comprise just one chain and have the round shape expected for an amorphous polymer, the height equals almost always the radius. With this in mind, a new attempt has been done to bring together experiment and computer simulations.

2 Simulation Method and Model

The calculations were performed using a modified version of EGO, a non-commercial molecular dynamics (MD) program [2]. The integration method of the Newtonian equations is done via a Verlet algorithm. All simulations are performed at isothermal conditions by coupling the system to an external heat bath [3]. The CH_2-groups are represented by a united atom model with a mass of 14 amu. The length of the investigated chain is 500 united atoms.

For the chain modelling we are using stretching, bending and dihedral as well as a nonbonded interactions between nonadjacent atoms. Specific

Springer Proceedings in Physics, Volume 85
Computer Simulation Studies in Condensed-Matter Physics XII
Eds.: D.P. Landau, S.P. Lewis and H.-B. Schüttler

details and more information will be given in a forthcoming publication [4]. For the nonbonded, long-range van der Waals interaction a cut-off of 8 Å is used. Stretching and bending potentials are similar to potentials used for polyethylene, the helicity reminds of crystalline polypropylene or polystyrene [5] but no emphasis is given to specifically simulate a special polymer like polyethylene.

The surface is given by a Lenard-Jones potential which represents a plain attractive surface.

$$W(r) = -W_0[(\frac{r_0}{r})^{12} + 2.0(\frac{r_0}{r})^6] \qquad W_0 = 10.0 \text{ kJmol}^{-1}, \quad r_0 = 15.0 \text{ Å}$$

The strength of this potential is of the same magnitude as the nonbondend potential, which also leaves room for rivalry of chain-chain interactions versus adsorption.

The simulations were performed according to the following scheme:

Starting with a perfectly alligned all-trans conformation (Fig. 1a) and a random initial velocity distribution a collapse process has been followed. In a first attempt relatively strong potential parameters compared with realistic parameters were used to obtain clearer results for the later adsorption. In particular a 50 times stronger nonbonded potential was chosen. With this parameters a globular conformation, which resemble single chains drifting toward the surface in the experiments, could be generated. Under these conditions the collapse transition dominated by entropy were run for 300 ps.

Different temperatures and different random velocity distributions were used, obtaining globular conformations, which are clearly different from a random coil. An example of such a globular conformation run at 300 K is shown in Fig. 1b). This globule has a density of about 0.5 g cm^{-3} , which is about half of the density of crystalline polyethylene, polypropylene or polystyrene [5].

In a next step we used these globular conformations and added again a random initial velocity distribution, chose more realistic potentials and performed an equilibration process at different temperatures. The equilibration process of typically 10 to 50 ps is too short to make sure the mean square displacement of the center of mass reaches the magnitude of the radius of gyration. Hence, the systems are far from equilibrium. However, the electrospray experiment might also show non-equilibrium effects.

The equilibration process turns into an adsorption process when the surface chain interaction starts to act. This is realized by setting up the surface at a 5 Å distance to the nearest atom of the coil at a particular given moment during the simulation.

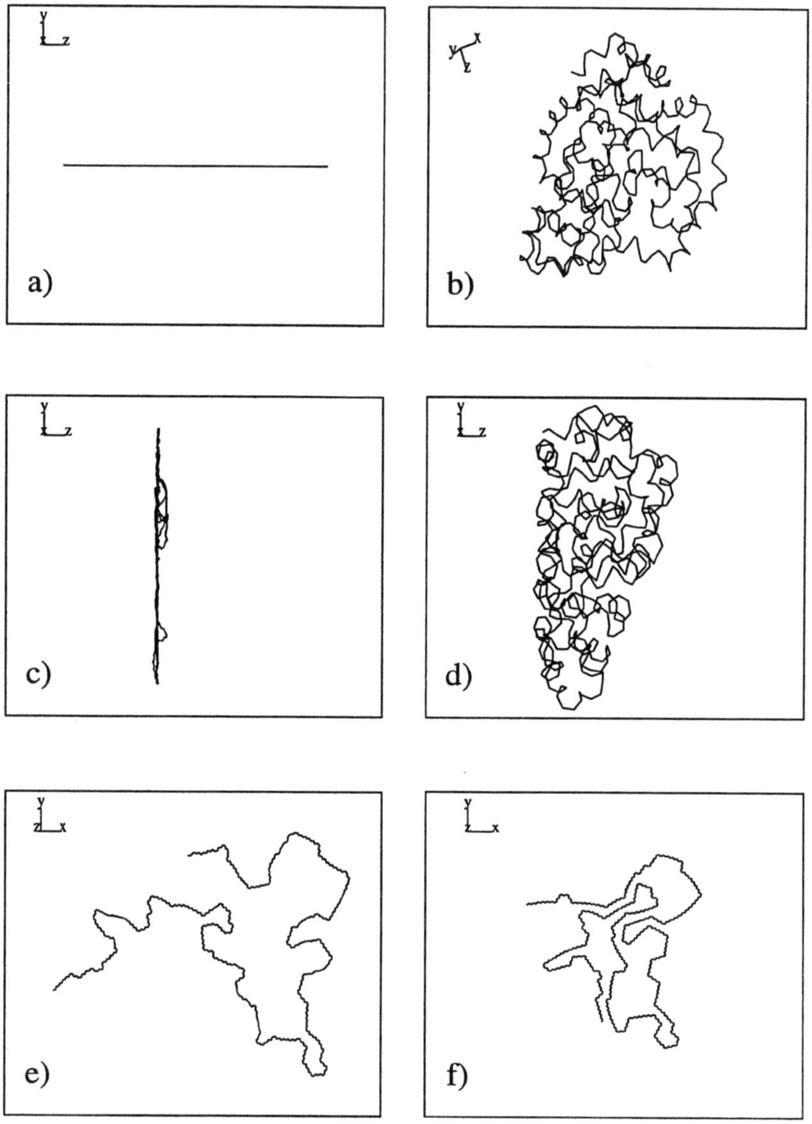

Fig. 1. a) Perfectly alligned all-trans conformation at the start; b) Globular conformation after a collapse process of 100 ps at 300 K; c) Pancake conformation after an adsorption process lasting 150 ps at 300 K; d) Globule conformation after an adsorption process lasting 150 ps at 300 K with stronger chain potentials; e) crossover-free conformation at 1250 K after 1100 ps; f) crystall-like conformation in two dimensions

3 Preliminary Results and Discussion

The adsorption process ends up with a pancake conformation, showing one or more layers of crossing portions of the chain. This is shown in Fig. 1c) for a temperature of 300 K after 150 ps of the adsorption process. Using strong chain potentials or alternatively a weaker adsorption potential, the globule conformation retains its shape, which resembles the semispherical shape seen in the experiment mentioned above. An example for such a conformation obtained by running the adsorption process with the potential strength of the collaps transition is depicted in Fig. 1d). At low temperatures, the shape and density do not considerably depend on how long the simulations are lasting.

Chain dynamics at higher temperatures make the crossovers in the pancake conformation disappear. This is shown by Fig. 1e), which is gained after 1100 ps performed at a temperature of 1250 K. Keeping the simulations running at much lower temperatures, for instance 100 K, these crossover-free conformations turn into crystal-like conformations. This crystallisation takes place in two dimensions. Fig. 1f) shows the conformation at 100 K, 400 ps after the one depicted in Fig. 1e). Clear signs of parallizing of chain can be seen. Further investigations are in progress.

4 Acknowledgement

This work was supported by the Division of Material Research, National Science Foundation, Polymers Program, Grant No. DMR-9703692 and the Division of Material Sciences, Office of Basic Energy Sciences, U.S. Department of Energy at Oke Ridge National Laboratory, managed by Lockheed Martin Energy Research Corporation for the U.S. Department of Energy, under contract number DE-AC05-96OR22464. One of us was supported by the Deutsche Forschungsgemeinschaft (DFG).

References

1. Festag, R., Alexandratos, S. D., Cook, K. D., Joy, D. C., Annis, B., Wunderlich, B. (1997) Single- and Few-Chain Polystyrene Particles by Electrospray. Macromulecules **30**, 6238-6242
2. Eichinger, M., Grubmüller, H., Heller, H., Tavan P., (1997) FAMUSAMM: An Algorithm for Rapid Evaluation of Electrostatic Interactions in Molecular Dynamics Simulations. J. Comp. Chem. **18**, 1729-1749
3. Berendsen, H. J. C., Postma, J. P. M., van Gunsteren, W. F., DiNola, A., Haak, J. R. (1984) Molecular dynamics with coupling to an external bath. J. Chem. Phys. **81** 3684-3690
4. Michel, A., Kreitmeier, S., Wunderlich, B., in preparation
5. Wunderlich, B., (1973) Macromolecular Physics **1**, Academic Press, New York London

Springer Proceedings in Physics

1 *Fluctuations and Sensitivity in Nonequilibrium Systems*
Editors: W. Horsthemke and D. K. Kondepudi
2 *EXAFS and Near Edge Structure III*
Editors: K. O. Hodgson, B. Hedman, and J. E. Penner-Hahn
3 *Nonlinear Phenomena in Physics*
Editor: F. Claro
4 *Time-Resolved Vibrational Spectroscopy*
Editors: A. Laubereau and M. Stockburger
5 *Physics of Finely Divided Matter*
Editors: N. Boccara and M. Daoud
6 *Aerogels* Editor: J. Fricke
7 *Nonlinear Optics: Materials and Devices*
Editors: C. Flytzanis and J. L. Oudar
8 *Optical Bistability III*
Editors: H. M. Gibbs, P. Mandel, N. Peyghambarian, and S. D. Smith
9 *Ion Formation from Organic Solids (IFOS III)*
Editor: A. Benninghoven
10 *Atomic Transport and Defects in Metals by Neutron Scattering*
Editors: C. Janot, W. Petry, D. Richter, and T. Springer
11 *Biophysical Effects of Steady Magnetic Fields*
Editors: G. Maret, I. Kiepenheuer, and N. Boccara
12 *Quantum Optics IV*
Editors: J. D. Harvey and D. F. Walls
13 *The Physics and Fabrication of Microstructures and Microdevices*
Editors: M. J. Kelly and C. Weisbuch
14 *Magnetic Properties of Low-Dimensional Systems*
Editors: L. M. Falicov and J. L. Morán-López
15 *Gas Flow and Chemical Lasers*
Editor: S. Rosenwaks
16 *Photons and Continuum States of Atoms and Molecules*
Editors: N. K. Rahman, C. Guidotti, and M. Allegrini
17 *Quantum Aspects of Molecular Motions in Solids*
Editors: A. Heidemann, A. Magerl, M. Prager, D. Richter, and T. Springer
18 *Electro-optic and Photorefractive Materials*
Editor: P. Günter
19 *Lasers and Synergetics*
Editors: R. Graham and A. Wunderlin
20 *Primary Processes in Photobiology*
Editor: T. Kobayashi
21 *Physics of Amphiphilic Layers*
Editors: J. Meunier, D. Langevin, and N. Boccara
22 *Semiconductor Interfaces: Formation and Properties*
Editors: G. Le Lay, J. Derrien, and N. Boccara
23 *Magnetic Excitations and Fluctuations II*
Editors: U. Balucani, S. W. Lovesey, M. G. Rasetti, and V. Tognetti
24 *Recent Topics in Theoretical Physics*
Editor: H. Takayama
25 *Excitons in Confined Systems*
Editors: R. Del Sole, A. D'Andrea, and A. Lapiccirella
26 *The Elementary Structure of Matter*
Editors: J.-M. Richard, E. Aslanides, and N. Boccara
27 *Competing Interactions and Microstructures: Statics and Dynamics*
Editors: R. LeSar, A. Bishop, and R. Heffner
28 *Anderson Localization*
Editors: T. Ando and H. Fukuyama
29 *Polymer Motion in Dense Systems*
Editors: D. Richter and T. Springer
30 *Short-Wavelength Lasers and Their Applications*
Editor: C. Yamanaka
31 *Quantum String Theory*
Editors: N. Kawamoto and T. Kugo
32 *Universalities in Condensed-Matter*
Editors: R. Jullien, L. Peliti, R. Rammal, and N. Boccara
33 *Computer Simulation Studies in Condensed-Matter Physics: Recent Developments*
Editors: D. P. Landau, K. K. Mon, and H.-B. Schüttler
34 *Amorphous and Crystalline Silicon Carbide and Related Materials*
Editors: G. L. Harris and C. Y.-W. Yang
35 *Polycrystalline Semiconductors: Grain Boundaries and Interfaces*
Editors: H. J. Möller, H. P. Strunk, and J. H. Werner
36 *Nonlinear Optics of Organics and Semiconductors*
Editor: T. Kobayashi
37 *Dynamics of Disordered Materials*
Editors: D. Richter, A. J. Dianoux, W. Petry, and J. Teixeira
38 *Electroluminescence*
Editors: S. Shionoya and H. Kobayashi
39 *Disorder and Nonlinearity*
Editors: A. R. Bishop, D. K. Campbell, and S. Pnevmatikos
40 *Static and Dynamic Properties of Liquids*
Editors: M. Davidovič and A. K. Soper
41 *Quantum Optics V*
Editors: J. D. Harvey and D. F. Walls
42 *Molecular Basis of Polymer Networks*
Editors: A. Baumgärtner and C. E. Picot
43 *Amorphous and Crystalline Silicon Carbide II: Recent Developments*
Editors: M. M. Rahman, C. Y.-W. Yang, and G. L. Harris
44 *Optical Fiber Sensors*
Editors: H. J. Arditty, J. P. Dakin, and R. Th. Kersten
45 *Computer Simulation Studies in Condensed-Matter Physics II: New Directions*
Editors: D. P. Landau, K. K. Mon, and H.-B. Schüttler
46 *Cellular Automata and Modeling of Complex Physical Systems*
Editors: P. Manneville, N. Boccara, G. Y. Vichniac, and R. Bidaux
47 *Number Theory and Physics*
Editors: J.-M. Luck, P. Moussa, and M. Waldschmidt

MIX
Papier aus verantwortungsvollen Quellen
Paper from responsible sources
FSC® C105338

If you have any concerns about our products,
you can contact us on
ProductSafety@springernature.com

In case Publisher is established outside the EU,
the EU authorized representative is:
Springer Nature Customer Service Center GmbH
Europaplatz 3, 69115 Heidelberg, Germany

Printed by Libri Plureos GmbH
in Hamburg, Germany